AL-BATTANI SHIELD

COUNTERACTING GLOBAL WARMING: A NEW APPROACH

INAYATULLAH IBRAHIM LALANI

iUNIVERSE, INC.
NEW YORK   BLOOMINGTON

Al-Battani Shield
Counteracting Global Warming: A New Approach

iUniverse books may be ordered through booksellers or by contacting:

iUniverse
1663 Liberty Drive
Bloomington, IN 47403
www.iuniverse.com
1-800-Authors (1-800-288-4677)

Because of the dynamic nature of the Internet, any Web addresses or links contained in this book may have changed since publication and may no longer be valid.

ISBN: 978-1-4401-8002-6 (sc)
ISBN: 978-1-4401-8004-0 (dj)
ISBN: 978-1-4401-8003-3 (ebk)

Printed in the United States of America

iUniverse rev. date: 10/22/2009

Dedicated to the Loving Memory of My Parents

Dr. Ibrahim K. Lalani

And

Mrs. Nabat Khanum Lalani

Whose boundless love and unconditional dedication to the
welfare of their
children – exemplified by the innumerable and extraordinary
sacrifices they
unhesitatingly made – still inspire us and enrich,
animate, and illuminate our lives.

FORWARD

Who, but a learned man with driving interest in the well being of all mankind, could have described for us the ideas contained herein. His careful painting of many factors that determine climate, pointing out that although considerable knowledge has been gained about the climate, there is much more that is unclear and unknown. This is an evenhanded interesting portrayal of what maybe looming in the future. He makes the case for a solution that would not deter the poor of the earth from climbing to a higher standard of living nor reduce the present standards that many already have and enjoy. The solution envisioned is a complex technical solution involving the construction of a daunting structure in space that would require enormous expenditures to construct. There is no solution that is both inexpensive and without detrimental effects on the poor. The solution offered in this work has important features. The entire effect produced by the shield could, if needed, be quickly halted and the program discontinued. That the average citizen of the world would hardly notice that the shield is in place and functioning.

Gordon G. Johnson Ph.D.
Professor of Mathematics
Houston, Texas
September, 2009

PREFACE

Artists, it is often asserted, are born and not made. Scientists are, on the other hand, made – more often self made – and not necessarily born, to be innovators or trailblazers. Once past the secondary education but especially at the doctoral level, ninety percent of knowledge that constitutes the expertise of a specialist is derived from self study and not by didactic indoctrination.

A Scientist must study, reflect, analyze, experiment, explore, question and sometimes boldly hypothesize. He must accept rejection, failure and sometimes even ridicule with equanimity because it is the establishment of truth about the nature of 'Nature' that is paramount to his quests; self assertion, self aggrandizement of self promotion are not part of his credo.

I enthusiastically join my readers, reviewers and critics in extolling the virtues of formal expertise, backed up with academic credentials. But often the definition of expertise tends to get too narrowly circumscribed and seems likely to cruelly stifle the scope of a scientist's inquiry. The society and civilization will be the poorer if 'institutionalized science', commissioned by the ivory tower types, is allowed to be the only game in town. And I do not speak out of disdain for recognized authorities; they could not have earned their places without talent, dedication, and hard work.

The scientific community is like a musical ensemble. Not every member of an orchestra can play every instrument, or even any other instrument than his own, but every musician intuitively understands and is able to follow a musical note emanating from whatever instrument being played on the conductor's cue.

This work does not claim to be of technical nature but is rather, largely, in the genre of popular science. That may partially relieve the writer of stringent constraints of academic credentials or formal expertise. I have freely exercised a privilege to *ad lib* certain well-established facts without rigorous and specific documentation since, for the readership to which this book is directed, such would be quite superfluous, besides being tedious and time consuming for this writer. If a generous allowance is not made for such license, how could William Calvin, a neurophysiologist, be such an effective exponent of climate science? Or how could Richard Harris of National Public Radio pontificate on so many varied fields of science? And how could Herbert George Wells – a non-scientist – have inspired and guided whole generations of *avant gard* scientists who changed the face of physics in the first half of the twentieth century?

And so I make no apology for not being a *professional* physicist or a mathematician. I am science indoctrinated; believe in scientific inquiry, am passionately committed to scientific method, and take a lively interest in all branches of the physical sciences, especially physics and astronomy. I read physics textbooks for relaxation. Above all, I am a fanatic about my (and everyone else's) prerogative to indulge in scientific inquiry – and speculation – regardless of 'official' *credentials.*

It is said that the practice of medicine is 80 percent mumbo-jumbo and 20 percent science and that the art of practicing clinical medicine is in balancing the two. Not so with me. I practiced medicine for thirty-five years only in so far as science dictated my conclusions and my decisions, yet never yielded to the voodoo aspect of medicine even while completely surrounded by the culture of shamanism. I think keeping in constant contact with the physical sciences during all those years made that kind of '*elitism*' possible.

This is my first venture in non-fiction creative writing of any magnitude and I have deliberately dispensed with many formalities peculiar to this vocation. Critics may find omissions of exhaustive and systematic annotations and conscientious cataloguing of bibliography less than welcome, but I hesitate to leave an impression of originality when the ideas presented are not my own. While studiously avoiding conscious plagiarism, I may have unconsciously quoted, without permission or attribution, some clever and particularly apt phrases not of my own coinage when I encountered them but could not remember where I did so. Attribution is provided where appropriate. When I do provide attribution, it is more likely from my memory and likely to be in error because I do not have the discipline to keep a detailed and methodical 'writer's journal'. I claim no originality in the work that follows save the concept of 'Heliosynchronous' trajectory (itself only a quasi scientific construct in celestial mechanics) and its application in the solution to the problem of global warming. All else is derived from my perusal of encyclopedias and some seminal works such as Richard Rhodes' "The Making of the Atomic Bomb," "Dark Sun," "Source Book of Physics," "Source Book of Chemistry," various works of Duncan Steel, and assorted textbooks and history books of science and astronomy and the monumental "An Inconvenient Truth," by Al Gore. Some ideas – such as the concept of a limited biosphere being the sole focus of climate change to the exclusion of the much more extensive remaining layers of the planet and the pitfalls of that narrow preoccupation are to be credited to Robert Claiborne's very readable and delightful book, "Climate, Man and History." A special gratitude is owed to William H. Calvin whose landmark 1998 article in *Atlantic Monthly*[1] about the great climate flip flops really animated my curiosity and set me on the road to creativity, as exemplified by this work.

---

[1]   See Bibliography

Formulating a coherent definition, or at least the descriptive outline, of the trajectory of an extraterrestrial barrier between our planet and our Sun has been a cruel struggle. Lacking formal, higher level education in mathematics made the struggle more agonizing. Constant perusal of quasi-technical periodicals and online resources such as Space.com, 'Sky and Telescope' website and broadcast sources such as National Public Radio and BBC World searching for clues understandable to a non-mathematician and *dilettante* physicist did pay off. As a result, it did become possible to enlarge on the subject (chapter 11). Even so, the ideas presented in that chapter, which enunciate the *raison d'être* for this endeavor, will not see the light of the day in the scientific-technical world without a generous preparedness of the venerable specialists in the field of climatology to meet me halfway.

My motivation is not to make money or win a prize but only to place the central idea behind this work before the leaders, scientists, economists, policy makers and anyone among the general public with interest in the fate of this planet. I am therefore asking for the readers' indulgence and request forgiveness for literary, scientific, and any other transgressions!

Like any other creative enterprise, writing a book is not possible without much help from loyal family members, friends and supporters; I would be amiss if I did not specifically identify a few individuals to whom I am particularly grateful. My wife Salma provided ample moral support and some help also in critiquing the book from the point of view of readability. Dr. Gordon Johnson, Professor of Mathematics at the University of Houston, gave generously of his time in reviewing earlier manuscripts and pointed out myriad inaccuracies and omissions that would have otherwise tarnished this work.

Veteran Editor Ron Kenner (www.RKedit.com) deserves special thanks for a very professional job of editing a very imperfect manuscript and for bringing it up to a publishable standard. In addition to the outstanding quality of his editing,

I am grateful for his many suggestions of substantive nature that added luster to this book.

And finally, I want to thank my publishers, iUniverse for their turnkey job as they saw the completion of their publishing assignment with competence and professionalism.

CONTENTS

CHAPTER 1

THE CONUNDRUM

During the last couple of decades, and with increasing intensity with each passing year, the issue of global climate change has dominated the public debate, not only in America but also internationally. When a grandee of the stature and eloquence of the former American Vice President – a Nobel Laureate – Al Gore makes the issue the banner of his crusade, humanity is compelled to listen and reflect; I listened and I reflected before sitting down at my computer to compose this work. This book seeks to inject one small item for consideration by those who, hopefully, in the coming years, would move and shake things in order to fix the 'problem' we all collectively face.

I placed the sobriquet 'problem' in quotation marks with due deliberation. In our lifetime and in those of many earlier generations, we have been gravely warned by our leaders – both current and erstwhile and by armies of sundry sages of the day – about many existential threats facing mankind that call for unprecedented alarm, exquisite preparedness, and supreme sacrifices to ward off the demise of civilization, or, at any rate, of Western civilization. I would, sometimes hesitatingly, demur, wondering if the threat of global climate change is not likewise destined to fall in the category of those threats that somehow harmlessly fizzled without a bang and are now little remembered and even less talked about. But then, I could be wrong. If I was as confident about my reservations as Al Gore is sure about his conviction, the justification of my labor in writing this book would vanish just like the threat of, say, the 'evil' ideology of Bolshevism did toward the end of the last century.

So let us now proceed with the assumption that the global warming proponents are right and let us briefly visit what they are contending.

Since the times shrouded in the mists of very early antiquity and even for long before then, the *mean* or the *average* or the overall climate of the Earth (the so called 'global' climate) remained stable[2] because of the marvelous equilibrium of nature that balanced – while allowing rather huge and abrupt variations from year to year and decade to decade – the opposing forces that tend to push the climate too far in one direction or the other. In the pages that follow I will bust the concept of 'global climate' because this thing called global climate is not that at all. It is only the sum total of meteorological factors (temperatures, humidity, barometric pressure winds, precipitation etc.), pervading a very narrow band inhabited by living things. Let us, however, not be distracted by that heresy right now.

A unique – and uniquely mischievous –creature that we now recognize as "human" appeared on the world scene about a hundred thousand years ago, at first little distinguished from other creatures in its subservience to his environment, pitifully small in numbers and cutting an altogether pathetic figure. Surviving with exquisite cunning not generally possessed by any other living thing, he multiplied exponentially, both in numbers and in his mischief-making potential. Obsessed with his own comfort, security and dominance, man has also been plagued with

---

[2]   In an interesting – but highly technical – article, "Earth Rings for planetary environment control," Pearson et al claim that for 95 percent of its past the Earth's climate has been warmer than now and that we are currently in an Ice Age and coming out of it.

the pernicious diseases of acquisitiveness and tribalism; behaviors which have embroiled him in endless discord and conflict with nature at large, as well as with his own kind, as he started taking over the planet for his exclusive use and at the expense of all living things which he corralled in his own service.

Man's need for comfort, security and dominance – not just over nature and all other living creatures but over other humans – has required him to use *energy* in increasing quantities. His consumption of energy reached prodigious proportions after the onset of the industrial revolution which started in an area of the planet roughly extending within a radius of about one hundred miles around Amsterdam in the low countries of Northwestern Europe some 250 years ago. When the obvious and easily accessible sources of energy such as firewood, cow dung and small scale windmills and watermills no longer sufficed to meet his ever growing hunger for it, man began prospecting for, and mining, sources of energy several magnitudes greater than he had historically been satisfied with. One such source is the fossil fuel – coal and hydrocarbon products of mineral origin - that had remained locked up in the bowels of the earth for millennia and left undisturbed by any creature or any forces that nature cared to deploy.

The profligate and reckless squandering of these fossil fuel resources is thought to be behind the climate crisis which has us so worried.

When a hydrocarbon derived source of energy – petroleum and natural gas – is burned, it produces, among other byproducts, carbon dioxide (CO_2), releasing it into the atmosphere. Same thing with coal, with the added problem of other noxious products that pollute the air we breathe. Not that firewood and cow dung don't produce CO_2 upon combustion, they do. But this CO_2 is only a recycled product, not released *de novo* from sources that have hitherto remained sequestered by the nature.

With the huge new inflow of CO_2[3], derived from mineral sources, which hitherto had been a very inconsequential component of the earth's atmosphere at large, the balancing forces that maintain the CO_2 concentration within a low, relatively narrow range (the so–called CO_2, or more correctly, the carbon cycle, discussed elsewhere in this book) may be undermined. The scholars tell us that CO_2 levels in the Earth's atmosphere today are more than twice, thrice, or even four times what they were at the onset of the industrial revolution.[4]

The high concentration of atmospheric CO_2 impacts the workings of nature in significant ways, some salutary and some very deleterious. The salutary impact – which is often dismissed by the pundits as being of dubious or marginal value, concerns the very desirable increase in luxuriance of the forests. Higher concentration of atmospheric CO_2 stimulates growth of trees all over the earth. Our forests are, regrettably, also endangered by man's desire to exploit and establish dominance over nature. Anything that helps the trees resist man's depredations should be welcomed by our species.

Our discourse mainly concerns the deleterious effects of high levels of CO_2 and so let us return to that aspect of CO_2 excess.

Reduced to its simplest, CO_2 is a *'greenhouse gas'*. It is not the only greenhouse gas, or even the most important one at that. But it is the greenhouse gas that seems to be running away and the one for which man bears explicit and exclusive blame, namely his recklessly growing use of fossil fuels.

How do the greenhouse gases impact the climate? Here is the theory (or shall we call it a hypothesis?). Just as in a green house (or the 'hot house' of horticulturists) the glass roofs and walls trap the sun's heat within the confines of the enclosure, the greenhouse gases in the atmosphere trap the sun's heat underneath the upper

---

[3] J. Pearson J. et al quoted an estimate by Govindasamy and colleague that atmospheric CO_2 will double over the next century resulting in earth's warming by $1 - 4^0$ Celsius.

[4] See Bibliography (IPCC)

atmosphere, causing a rise in overall 'global' temperature – the much talked about *greenhouse phenomenon*. Now, it should be understood that this wicked phenomenon is confined to that portion of the atmosphere which is closest to the surface of the earth. But Isn't that the region we care most about? What does it matter to us if the drastic changes in the atmospheric conditions are confined, say, to the stratosphere above it or the ionosphere on the edges of space? The answer, at first blush, is 'none at all'. However, we need to understand that there is a great deal of interaction between different layers of our Earth's atmosphere and that most of these interactions are poorly understood.

Assuming then that there has been an unwholesome increase in the levels of atmospheric CO_2, and that, in consequence of the villainous greenhouse phenomenon attributable to the latter, there has been a measurable and significant global warming[5]. If this is rapidly accelerating as a result of the worldwide increase in the production and use of energy, what consequences do we have to worry about?

Climate scientists tell us that global warming – if it is allowed to grow unchecked – will have catastrophic, even apocalyptic, effects due to the melting of polar ice, to disappearing permafrost and the vanishing glaciers. The sea levels will rise, the low lying areas of earth (where much of the human populations live) will be inundated, there will be an increase in the frequency and intensity of calamitous weather phenomena such as hurricanes, tornadoes and other cyclonic activities. Many areas of Earth will be hit with droughts and famines, and pestilences and plagues may become pandemic. Poverty will spread as living standards fall and there will be wars (don't we already have them?)

---

[5]   According to the Intergovernmental Panel on Climate Change (IPCC), the mean global temperature has so far risen 0.74 ± 0.15^0 C in the last thirty years and can be expected to rise as much as an additional 4^0 C in the next 50 years even if the CO_2 emission is held at the current levels!

Paradoxically, at least in the shorter run, there might be a return of the Ice Age over certain areas of the Earth. But that's not the worst of it. At some time, when an undefined tipping point is reached, the global warming will spin completely out of control and will become irreversible. That would mean that life as we know it may become impossible to be sustained at the ungodly temperatures at which the earth's climate may eventually settle.

So the obvious solution to the problem of global warming is reversing the rate at which the whole of humanity uses fossil fuels.

So where is the conundrum? Let me show you. Far from achieving a reversal, or even arresting, the growth of the use of hydrocarbons, all indications point to just the reverse. The current 'crisis' is attributable to the high living standards enjoyed by only about 15 percent of humanity (comprising the affluent, industrialized world). Even in the unlikely event that the affluent countries agree to hold their living standards at the current level, there's no likelihood that the developing countries will accept the *status quo* and forego their aspirations for higher living standards similar to those long enjoyed by the more affluent societies. Such an achievement would raise the CO_2 production several fold, a sure guaranty of runaway global warming! There seems no surefire way of providing enough energy and the implicit high living standards for the affluent as well as the developing worlds without further huge growth in the burning of fossil fuels. Turning to alternative sources of energy – especially renewable sources such as solar, wind, tidal, etc., and utilizing non-renewable and controversial nuclear energy– these are laudable goals and should definitely be pursued. But they are hardly likely to address the underlying problem of controlling the use of coal, oil, and natural gas. So we may be condemned to a very sorry fate. That is the conundrum.

A conundrum calls for searching for solutions by *'thinking outside the box'*, to use the trendy new – and by now trite –cliché.

CHAPTER 2

WHAT IS CLIMATE?

The phrase 'Earth's Climate', or 'global climate', is used quite frivolously by most people. The climate, in reality and as it concerns humans and other living things, is the sum total of temperature, winds, barometric pressure and precipitation, etc., over a surprisingly narrow band (narrow in vertical sense) occupied by all living things. Since the ocean is nowhere more than seven miles deep and no life in any form is known to exist more than five miles above sea level, we are talking about a twelve mile wide band, at most, where those elements matter to us. We shall show that what happens below and above this band (loosely called the *biosphere,* the home of life*)* massively impacts on what happens to the climate within the biosphere, although these influences are neither linear and straightforward nor predictable. Nor are they even well understood. Without a doubt, they interact mutually in very complex ways. However, all that matters to us is what happens within the biosphere. I point this out to underscore our need to acknowledge that in our desire to control or modify or somehow influence what happens within the biosphere, we must, at all times, be cognizant of our limits concerning what happens either above it or below it. What we do within this biosphere, say, to reduce greenhouse gas emissions, may not produce the expected results; in fact our zealous efforts may backfire for reasons we don't completely understand. Measures taken to modulate the lower atmospheric conditions may alter the conditions in the upper atmosphere in ways that we may not like.

The climate of this home of life is the sum total of a great many factors that impact upon the final result in myriad ways; the impact of some factors is minimal, short-lived and easily reversible; for others, it is very profound; it may be far-reaching and, perhaps, irreversible. Some factors can be controlled or, at least, modulated by man. Others are completely beyond our ability to modulate.

The consequences of some human endeavors to modify these factors may be reasonably predictable, such as building a shelter to avoid excessive cold, winds and rain, albeit over a very limited area. When we get inside the shelter, we avoid getting wet, excessively chilled, and are spared unpleasant winds. When we drain the swamps, we change the eco-system (e.g., get rid of mosquitoes). The impact of other, larger, measures may not at all be what we intended because each change – whether spontaneous or induced by man – may set off secondary changes. The final outcome of those changes may not be what we had intended or foreseen. We don't always know or understand what secondary changes may follow and over what period of time when we execute a particular action to modulate the goings on in the home of life. This is the concept of *chain reaction*. These chain reactions may acquire lives of their own and the final outcome may not be what we set out to achieve. For these reasons, at this juncture we must continue to intensify scientific inquiries into the workings of nature and be very cautious, humble, and also philosophical, about what we can and cannot do.

In the next five chapters, I will try to explain and analyze the complex web of various factors that produce what we call 'global climate'. As noted earlier, 'global' is a vainglorious term for us living beings – but specifically us '*Homo sapiens*' because the *climate* that concerns us is not at all global. It only pertains to the twelve-mile wide band around our planet in which life currently thrives.

And a few words about the 'weather'. Weather is, as we all know, the set of meteorological conditions prevailing over a circumscribed area during a specified, short period of time; perhaps measured in a few days. It is a fleeting phenomenon that is quickly forgotten yet nevertheless dominates our immediate thoughts and keeps the TV meteorologists in business. Sure, we would like to know if it is going to rain on the Sunday we have set aside for a backyard barbecue; and we'd like to know how many snow days in a given season will force school pupils to postpone the end of the school year, and how many baseball games may have to be made up when they get rained out. And of course, we'd welcome information on the catalogue of broken bones from skidding on iced over pavements, on the heart attacks shoveling snow and on the sometimes fatal car accidents that alter lives for good. These matters are all very important to individual humans and to whole societies, but I mention these to warn you that I do not intend to address the issue of the *weather* – as opposed to the *climate* – any further in this book.

In subsequent chapters I put forward a novel proposal to employ our technical capabilities to prevent runaway global warming in such a way that no set of unintended consequences will cause us grief. I am not a mathematician and that is my greatest sorrow because, even though I am convinced about the validity and feasibility of the suggested project, the requisite mathematical arguments will have to await input from other professionals.

CHAPTER 3

A PRIMER OF CLIMATOLOGY
The Primary Building Blocks of Global Climate

The ultimate engine of global climate is the heating of our planet by the radiation from our Sun. Beyond this simple and rather obvious condition, a host of other factors interact with each other to produce the climate that prevails over our planet. Some of these factors are well understood, some are rather arcane, and still others have probably not even been thought of yet by human beings.

The basic factors that underlie the very agreeable global climate we humans have been so fortunate to inherit, the factors that make life not only possible but very enjoyable,[6] have not changed over millions of years. We need to understand these

---

[6] The singing and dancing of the unlettered, dirt poor and always threatened multitudes breaking out with reckless abandon in celebration of the harvest festivals of myriad old and new civilizations, long before we had cars, television, telephones, computers and airplanes or even light bulb, the intellectual and artistic achievements of the ancients and, above all, the irrepressible laughter and hope of our forebears who were so deprived by our modern day standards! That is what is meant by 'enjoyable' when we talk about the climate. Optimism, not gloom, has been the dominant human affect in the prehistory, unlike in our sullen modern, technologically advanced age.

very fundamental factors clearly before we explore other, more complex factors and phenomena. These factors are

(1) Earth's Intrinsic ('Native') Heat
(2) The rate of Solar Radiation
(3) Earth's distance from the Sun
(4) Length of Earth day and of the solar year
(5) Inclination of the earth's axis in relation to the plane of its orbit around the Sun *(The ecliptic)*

All these factors are of an astronomical nature.

As far as we know, no astronomical factors such as the influence of the Moon and other planets and the more distant celestial objects such as the stars, nebulae, constellations, comets, asteroids, etc. – nor such dramatic celestial phenomena as eclipses, the appearance of comets and novae, fireballs, meteorites and meteor showers etc., – have any measurable effect on the overall climate of our planet. Only those who subscribe to the mumbo-jumbo of astrology[7] allege that these factors are important and actually impact individual human life. No serious scientist is known to have delved on those factors to explain the global climate.

In trying to understand Earth's climate, it may be useful and interesting to think of our home as merely one planet out of about ten or so in the solar system and perhaps among hundreds, thousands or even millions of such celestial bodies in the universe. Admittedly, our home planet is *'blessed'* with some

---

[7]   Mesopotamians appear to have been the first to notice and name the twelve constellations of Zodiac through which the sun makes its annual sojourn; it was probably the Sumerians who cooked up the fancy tales about their influences on individual human destiny. The 'science' of astrology then spread to Egypt, next to Greece after which Alexander's invasions brought it to India. The culture of astrology seems to have died out in the land of its birth, robustly surviving only in India which is now the only place where a serious body of professional astrologers actually advises people about serious and fateful undertakings such as marriages.

unique characteristics that would be the *envy*[8] of all other known celestial bodies.

Let us call the above five factors the *prime factors,* or the main levers in the machinery of climate making. It is generally agreed that over the past several million years, maybe over several hundred million years, there has been no significant change in any of these factors, with the sole exception, perhaps, of the prime factor number two, namely the quantity of energy output of our Sun and the continental drift that resulted in the breakup of the supercontinent *Pangaea,* first into the Northern megacontinent of *Laurasia* and the southern *Gondwanaland* and later into the present land masses that we call continents. The significance of the former is debatable, but more about that later (see Appendix C). It's undeniable that rearranging the outlines of our continents has had a very profound effect on our climate. However, the reconfiguration of our continental land masses seems to have been essentially completed a couple of hundred million years ago and can safely be discounted in our calculations.

Even given that these five prime factors have been like the Rock of Gibraltar anchoring the fate of *Homo sapiens,* indeed of all living beings, there have undoubtedly been a number of severe upheavals in the Earth's climate. These range from ice ages lasting hundreds of thousands of years, enveloping variable, sometimes large areas of our planet, alternating with long periods of marked warming, perhaps much greater global warming than we are being warned about these days. Actually, there's little doubt that we are currently living in a blessed interlude ('interglacial' period) that started while we were still in the Paleolithic age, struggling to discover secrets of agriculture and domestication of animals and taking the first steps toward the technological age; an age that has brought us to this modern era where human existence is

---

8 Envy? By whom and to what purpose? Envy is a strictly a human emotion. The rock on Mars or methane geyser on Titan, a saturnine moon, would not really envy us!

potentially so much more enjoyable and hopeful. Actually, there's very little doubt that it was the receding glaciers and thawing of most of the planet that provided the greatest impetus for humans to blaze new trails,[9] both literally and figuratively.

Before we delve further into other possible factors (call them '*secondary*' and '*tertiary*' *factors*) that explain the epochal climate change for which there is unquestionable and detailed geological evidence, I would like to explain the *prime factors* in some detail.

Although, at the end of this discourse we may be forced to conclude that none of the five *prime factors* contributed to the giant upheavals in Earth's climate, we might still have to do something about some of them to stave off an unacceptable alteration of planetary climate because of our inability to modify (or even fully understand) the *secondary or tertiary factors* (described in later chapters) as the agents of change. On its face, this approach appears Quixotic, but let me finish my argument.

NATIVE HEAT OF EARTH

A most dramatic demonstration of the importance of Earth's internal heat unfolds in the colder climes during the winter for someone driving over an elevated highway ramp; the driver encounters fairly dry surface roadways and is fooled by it until

---

[9]   The known and named adventurers and explorers of the modern (historical) ages such as Marco Polo, Ibn Batuta, Columbus, Edmund Hillary, et al, cannot hold a candle to those intrepid explorers and sea-farers whose identity has been snuffed out in the mists of pre-history. The former we rightly celebrate – and lionize - in our history books, cinemas and literary works of fiction and non-fiction, but what about the latter? As they struck out over the mountains, through deserts and dense forests, across the mighty rivers and forbidding oceans as humanity overspread almost the whole planet against impossible odds, the challenges they faced and the hardships and danger they endured are the stuff of which supernatural legends are made and can hardly be matched by modern man with his marvelous technology to guide him, to protect, comfort, and reassure him.

he or she climbs up the ramp whose surface is slick with ice and sudden skids with the potential for serious accident. Thus at the entrance to those ramps the highway departments place warning signs: "Watch out for ice on the bridges." The reason is very simple. Surface roads are in contact with the earth whose heat warms the pavement through conduction and melts away whatever ice that may be forming. The elevated ramps are separated from the earth and, lacking contact with it, receive no benefit of its heat and thus are much colder. Ice thus forms on them. The intrinsic heat of the earth, in fact, is the *single most important* factor deciding the *base* climate.[10] Heat received from the Sun builds on the base temperature guaranteed by mother terra.

Spelunkers will tell you that the temperature within the caverns everywhere is a constant 56° Fahrenheit year round, regardless of what is going on outside the caverns. The miners who work deep underground will attest to the hellishly high temperatures at great depths. At depths greater than a few hundred feet, the mines must have a system of fans and air-conditioning to enable the miners to work. All these conditions demonstrate that there's a great deal of heat within the bowels of our mother planet.

A more persuasive evidence of the immense store of heat within the belly of the Earth is the phenomenon of geothermal energy. Most of us have experienced visits to 'warm' springs as well as 'hot' springs. These are found in almost every region of the earth, including the polar areas. The water issuing from many hot springs can be scalding and, in countries endowed with plenty of volcanism, they are of industrial significance. In Iceland and the Philippines, for instance, the energy of the superheated steam

---

[10]   This statement will perhaps be challenged and needs some elaboration. It is true that the amount of heat emanating from the interior of the earth is only about 1/20,000 that of the radiant heat received from the Sun, with a base temperature of about 56° prevailing just below the surface is the most relevant fact of our climate and that fact is entirely a consequence of the native heat of the planet.

gushing forth from the fissures in the earth is significant enough to be utilized for home heating and/or power generation.

Earth is a marvelous furnace which produces prodigious amount of heat in its center. This has been known to man for centuries though its causation remained a mystery. Even if the earth started out very very hot, its inevitable cooling over the millennia since the birth of the solar system (OK, since the *genesis,* if you prefer) ought to have depleted even the huge stores of heat with which our planet may have been initially endowed. This has evidently not happened. The natural conclusion must be that new heat is being generated inside the earth's core. There are no known chemical processes or fuels that could assure such immense heat production over millions of years. That enigma would have remained unsolved forever until man began to understand the phenomenon of radioactivity and nuclear fission, the same phenomenon that underlies the energy generation in the nuclear power plants, and, of course, the atomic bombs.[11] The radioactive fission on a large scale involving the fissionable U^{235} isotope (and, to a lesser extent, other radioactive elements

---

[11]   The nuclear furnace at the core of the earth, like the core of the nuclear power plant, conducts a controlled nuclear fission chain reaction; the reaction is controlled because the radioactive fuel rods consisting of the critical mass of low enriched uranium (i.e. 3.5% to 5% U^{235} isotope mixed with 95%-96.5% U^{238}) are kept apart from each other by the interposition of graphite rods (or their more modern substitutes) that serve as moderators ('tampers') of the chain reaction. An atomic bomb explosion, on the other hand, is an example of an uncontrolled nuclear chain reaction that, within milliseconds, consumes all the highly enriched (50% -90% $U^{235)}$ uranium (or plutonium), resulting in an enormous explosion from the release of immense quantities of energy. Of course the heat production in the Earth's core remains slow and naturally 'controlled'; instead of graphite rods, the non-fissile isotopes of uranium as well as many other elements serve as the tampers. Uranium is very plentiful on earth (especially near its core) and is the 14th most abundant element.

such as Thorium and fissionable isotopes of Potassium) is the principal source of the native geothermal energy.[12]

This reaction that generates such prodigious quantities of heat is a *nuclear* reaction engendered by *fission* of unstable radioactive isotopes, not a *thermonuclear* reaction resulting from the fusion of atoms (or their components). Fusion reaction can only be ignited when the prevailing temperature exceeds four and a half million degrees K. Such high temperatures on earth are only reached in the core of the exploding atomic bomb (of substantially greater yield than the 'baby' bombs that leveled Hiroshima and Nagasaki) and then only for a few microseconds! The weapons designers took advantage of this fleeting opportunity to ignite a fusion reaction between very light atoms (Hydrogen[3] or Lithium[6]) and we had the *Hydrogen Bomb*, a far more lethal weapon even than the atomic bomb.

What makes a star a star and not just an inert ball of gas is the presence of ongoing fusion reaction in the center of that blob. To initiate and maintain fusion reaction, an absolute requisite is the very high temperatures achieved in the core of the exploding atom bomb. Such temperatures are reached in the center of the stars, not by exploding atomic bombs but by the ungodly gravitational pressures resulting from the enormous mass of gases that constitute a star.

For the core temperature to reach four and a half million degrees, a ball of gas must be large enough to have the requisite pressure in its center. The great planer Jupiter, even with its

---

[12]   To assure scientific accuracy, it should be noted that our planet began with a huge store of heat inside its core, and that particular store of heat was acquired through the process of planetary accretion and its related phenomenon gravitational binding energy. The subject is quite esoteric but since there is no ongoing generation of heat in that department and since, in the absence of nuclear fission, most of that heat would have leached out the earth by now anyhow, we can safely disregard that particular source of heat.

considerable size, is not that big and has thus remained only a planet and not a star. Something well over ten times the mass of Jupiter would have sufficient mass and the requisite conditions in its center to ignite a sustained fusion (thermonuclear) reaction. Such midget stars, however, don't do too well and fizzle out in a short time (only a few million years).

Once a fusion reaction is initiated, the resulting super high temperatures ensure ongoing fusion processes and the production of huge amounts of energy then radiated in the form of light, heat, X radiation, etc.

If there were not any continuous production of heat in the earth's core,[13] the globe would now be much colder, possibly close to absolute zero (i.e. minus 273° K) and all the solar radiation in the world could not make this planet habitable. Our climate is so equitable because it is superimposed over a pleasant base of 56° Fahrenheit temperature at the surface of the earth, reached because the heat at the core of the planet reaches the surface by the phenomenon of *conduction* and not by *convection* or *radiation*.

It may be of little interest to the casual reader as to what exactly nuclear fission is and the fascinating history behind its discovery in the late 1930s. However, the story of atomic fission is so exciting and awe-inspiring that a brief recounting seems relevant. To minimize digression, the topic is discussed in a separate appendix (Appendix B).

---

[13]   An unattributed item was published in January 2009 issue of Scientific American reproducing the scientific discourse of a hundred years ago (1909). It contained a 'hypothesis' of causation of earthquakes. The author contended that the earthquakes resulted from deformation of the earth's crust as it cooled, much the same way that railroad tracks buckle in presence of extremely cold temperatures. Of course, the railroad engineers long ago solved that problem by leaving gaps between segments of tracks at regular intervals permitting both shortening from cold and lengthening from heat without deformation or cracking..

SOLAR RADIATION:

In the eyes of the astronomers, our sun is an *ordinary* star. To us, there is nothing ordinary about this life-giving ball of gas that continuously produces prodigious – and constant – amounts of energy which it emits into space. It has been doing so for more than four and a half *billion* years and is expected to continue doing so for several billion more. We, on Earth, receive only an infinitesimal fraction of the radiant energy put out by the Sun and yet, from this small ration spring, ultimately, all of the energy using processes on earth, including the weather, life, and photosynthesis.

The word *insolation* means the receipt of sun's radiation incident upon whatever celestial body receives it; in this case, of course, our planet Earth.

One more obvious fact is that insolation is greatest when the sun's radiation falls on the receiving surface at a right angle because a given amount of energy is received by the smallest possible surface area. As the angle of incidence grows more acute, the unit energy is spread out over an ever larger surface area; hence the rate of insolation decreases. At or near zero degree angles, the insolation tends to approach a zero value. Thus solar radiation is greatest when the Sun is directly overhead (possible only in the tropical regions of Earth, i.e. between 23.5 degrees North and South). The insolation at the poles would be zero at all times but for the axial tilt of Earth's rotation. This enables each pole, alternately, to receive some insolation. At the height of the summer (the summer *solstice* in the Northern hemisphere and the winter solstice in the Southern hemisphere), the Sun is as high as 23.5 degrees above the horizon, albeit only for one day. This is one of the myriad devices to moderate the global climate, even in the Polar Regions. Another one, of course, is the roughly spherical shape of the globe.

Remarkably, the rate of energy emission from the sun does not seem to have varied much over the eons. Thus insolation remains a fixed quantity, or so it is currently believed. Put another way, the Sun is a 'non-variable' – or fixed – star. Not all stars share this characteristic of our Sun; in fact most stars are variable to some degree, and some extreme examples vary their energy output by a factor of one hundred in a matter of days or even hours. This kind of variability, where the rhythm of changing output is predictable, makes a star a *regular variable* star. There are *irregular variables* whose changes in energy output are completely erratic and unpredictable. The Sun, however, faithfully emits the same amount of energy hour after hour, day after day, millennium after millennium. It is an extremely stable dude![14]

I am likely to be vigorously challenged about the above assertion by a sizeable segment of a community of serious, mainstream scientists. They will immediately point to the recurring phenomenon of eleven year-long sunspot cycles and the effects of solar maxima and minima (during which the number of sunspots are at the maximum or minimum, respectively) and their correlation with global climate. The subject is too large to be addressed here but the arguments against the role of Sun's variability in changing climate are recounted in Appendix C.

PROXIMITY TO (DISTANCE FROM) THE SUN

The average distance between our Earth and the Sun is expressed as one *'astronomical unit'*, approximately 93 million miles. Since the orbit of the Earth is not perfectly circular but elliptical, the Earth's distance from the Sun varies, by less than three million miles, between the nearest and the farthest points.

---

[14]   There are some staunch advocates of the view that our sun, like most stars in the universe, is indeed a variable star whose rate of energy output varies significantly over long periods and that many, if not all, periods of glaciations and interglacial phases are primarily due to such oscillations. Those who hold such a view are in minority. One argument against solar variability is that if that were the case, the length of the cycle of variability would have to be uncharacteristically long. We know of no known variable star whose period exceeds a few months.

Earth reaches its *perihelion* (the point nearest to the Sun) sometime in January and the *aphelion* (the point farthest from the Sun) in July. However, this situation will be reversed in a few thousand years because of a phenomenon called precession of perihelion. Briefly, the time spot in the year (i.e., alignment with a particular region of the Zodiac) when Earth reaches its perihelion (the same as *perigee)* is slowly and uniformly slipping[15] counter clockwise; thus in about ten thousand years the perihelion will occur in July and the aphelion in January. However, that reversal will not alter the global climate perceptibly or measurably. The effect of Earth's distance from Sun over the degree of insolation is quite negligible, less than one percent; much less significant than the prime factor number 4, discussed below.

The degree of insolation is inversely proportional to the square of the distance from the Sun. A planet twice as far from the Sun as our Earth will receive one quarter as much insolation. One orbiting at half the distance will receive four times as much insolation as does our mother planet.

But as we shall elaborate later in this chapter, the distance between a star and its planet is only one factor affecting the planet's climate. It is possible for a planet further from the Sun to be warmer than another one which is actually closer (and hence receiving greater insolation). Mercury is substantially closer to Sun than Venus, yet is not as hot as Venus. The little planet has lost all its intrinsic heat and is not thought to be housing the nuclear furnace inside its interior. If there is no intrinsic heat,

---

[15]   The rates of precession of the equinoxes and of the slippage of aphelion are so nearly identical (1^0 every 72 years) that many ancients assumed them to be the same – a classical example of Ockham's Razor. In fact, starting two generations before the birth of Al-Battani - 776 A.D. –and for 150 years after his death, a raging controversy persisted over this subject. The elliptical shape of the Earth's orbit was well known to the ancients as attested by the new kind of armillary sphere, designed by that skilled instrument maker Al-Battani, who called it al bayda (Arabic for the 'egg'). It was egg-shaped, (an oblate spheroid), to underline the eccentricity of earth's orbit.

there can be no volcanism. And there is absolutely no evidence of volcanic activity on Mercury.

Extremely dense Venusian atmosphere consisting largely of carbon dioxide (laced with sulphuric acid thrown in good measure) – our villain *de jure* greenhouse gas – is partly responsible for this anomaly. Of course, like Earth, Venus also has vigorous volcanic activity; this suggests an immense store of native heat inside that planet too, arising out of the same mechanisms, namely planetary accretion and, almost certainly, nuclear fission.

Another explanation (for this paradox of the more distant planet being hotter than the one circling closer to the Sun) may result from the common understanding of what constitutes warm and hot. Scientific measurements of temperatures at various altitudes of Venus' atmosphere will invariably produce some strange results. These readings may not jibe with our preconceived notions. But, in general, Venus is hellishly hot and Mercury (only its surface, for there is no atmosphere in which to measure and record temperatures!) somewhat less so.

This brings us to another, somewhat philosophical point. The variations in our planet's temperatures that we intermittently call ice age or the warm, interglacial age, although great in human experience, are not, in nature's perspective, all that great. In the human perspective, we are concerned only with the temperatures prevailing at the altitudes at which we live, viz. at sea level or a few hundred (or thousand) meters above or below that level. Yet, as suggested, what appear to us as extremes in temperatures are not that extreme if you take a wider view of the nature. Really cold would be less than 300 degrees below zero (as occurs at night, on Pluto, for instance) and really hot would be several million degrees, as found in the centers of stars. By contrast, because of the fragility of life, especially the higher forms of it, and because the freezing point of water is so critical to many of the weather phenomena on earth and to the biological processes, the range of temperatures which concern us is indeed narrow. Yet from nature's wider perspective the deepest of ice ages was not much colder than the current, rather pleasant global climate conditions.

Things could get much nastier, however. Venus, a planet much like ours in many ways (about the same size and not much closer to the Sun than our Mother Earth) should not be a whole lot warmer than our planet if everything else were equal. But everything else (by which I mean the secondary and tertiary factor I mentioned earlier and which I will discuss in much greater detail shortly) is not equal; thus the surface temperatures on Venus regularly exceed 800 degrees F instead of the anticipated, maybe, 130 F degrees at mid-day near the Venusian tropical zones and a very pleasant 80 F degrees in near polar regions of that planet if that planet was like ours in more ways.

LENGTH OF DAY AND OF YEAR

An important reason why the mean atmospheric temperatures on the Earth's surface remain within the relatively narrow, life-sustaining range, is that our planet rotates on its axis once approximately every 24 hours, not as short as the Jovian day of 8-10 hours but not anywhere as long as the Venusian day of more than 5,800 hours! Consequently, every part of our globe is warmed by the Sun during the day and then gets a chance reasonably soon thereafter to cool down somewhat, so that it does not overheat. This twenty-four hour period, the most optimal length for the day is so normal and predictable that we take it for granted and never think what would happen if the day was shorter or longer. We can see how unpleasant things would be if the rate of Earth's rotation decreased or, especially if it increased, significantly. A somewhat related factor is the Earth's mass and how that affects the protective atmospheric blanket; this largely determines how the force of gravity would come into play – in a very unpleasant way – if the earth's rotation speeded up. With the earth spinning more rapidly, we would weigh less but that would be about the only thing we'd like. Gradual loss of our atmosphere would be something we could not countenance, yet that would be a sure negative outcome of the faster rotation of our earth. Also, the ocean tides would be higher, forcing us to give up some of coastal areas that are currently dry and habitable.

And poor Venus! Not only is it shrouded in a very thick layer of atmosphere made up, largely, of carbon dioxide but it also has that very long day (243 earth days, longer than its year of 224 earth days!). This unbearably long day only compounds its misery[16].

The orbital period of the earth around the Sun, that is, the length of the solar year is also an optimal quantity contributing to the Earth's well being and, especially, it being so hospitable to life. This is mostly because the length of the year controls the length of the seasons. The four seasons, which so animate the poets because they enliven (literally) our human condition, are not caused by the length of the orbital period or the orbit's eccentricity. They are caused by the inclination of Earth's axis of rotation in relation to the plane of its orbit. Length of the seasons, roughly every three month period, is definitely the function of the length of the year. If the year were much longer, the seasons would be correspondingly longer, the hurricanes, typhoons, and cyclones would be much more frequent and more violent, the winters much colder and longer, and so on. Life, at least some forms of it, would definitely adapt to longer – or shorter – years and longer or shorter seasons, but we humans would definitely be more miserable because we are so spoiled. So that brings us to the main discussion of the last of the five prime factors, namely the genesis of seasons caused by the inclination of Earth's axis.

---

[16] What 'misery'? Venus, of course feels neither misery nor ecstasy! It is miserable out there only from our anthropocentric, narrow-minded point of view.

INCLINATION OF EARTH'S AXIS AND THE GENESIS OF THE SEASONS.

The North Pole of the Earth stays pointed toward the Pole star. (The hallowed place of Polaris is not written in stone. Several different stars take turns as the Pole star over a period of 26,800 years, but that is a separate matter) as it sojourns around the Sun in the course of the year. The pole star for Earth (currently *alpha ursae minoris)* is not the Pole Star of our Sun or any other planets in our solar system. That is because the axes of rotations of the Sun and the Earth – and other planets - are not parallel. The axial tilt of the Earth, about 23.35 degrees relative to the plane of its orbit (also called the *ecliptic),* results in several interesting things. For half of the year, the North and South poles of our planet alternately face toward or away from the Sun, causing varying lengths of day and night and also the seasons. Briefly stated, at the summer and winter *solstices,*[17] one or the other pole is most tilted toward – or away – from the Sun. And at the spring and autumnal *equinoxes*, at roughly midway points between the solstices, the poles are pointed neither toward nor away from the Sun. Thus, albeit for just one day (March 21 and September 23), daylight and night periods are exactly equal all over the planet. In the phenomenon underlying changing seasons, the hemisphere pointed toward the Sun for half the year receives more insolation (first increasing and later decreasing) than the other hemisphere which receives first decreasing and then an increasing amount of solar radiation. The first half of that warmer half year is the spring; giving way, at solstice, to summer. During the summer,

---

[17]   Solstices are two points, 180 degrees (i.e. half a year) apart from each other, where the ecliptic intersects the earth's equatorial plane. 'Ecliptic' is the plane of Earth's orbit around the sun; it does not coincide with either the plane of earth's equator or of the plane of moon's orbit. The twelve constellations of the Zodiac are in the plane of Earth's orbit, i.e., in the ecliptic. Separately, the sidereal year is about four minutes longer than the solar year; these minutiae are not relevant to the subject on hand.

the actual amount of insolation is steadily decreasing but the accumulated heat causes the weather to get warmer despite the steadily shrinking day (and decreasing insolation).

The peak of summer temperatures on land is reached some five weeks after the summer solstice. The length of the day is steadily shrinking during the summer and grows steadily longer during winter. So autumn lays groundwork for the winter and the spring lays the groundwork for summer. The net effect of this seasonality is overall *moderation* of the global climate. No part of earth overheats and, with the exception of the Polar Regions, no area freezes. Even the Polar Regions escape from what would otherwise be even more frigid conditions.

To say that the seasonality is life-giving is too human centered a view-point; rather, life adapted itself to this phenomenon.[18]

Seasonality also facilitates other phenomena (discussed under the secondary and tertiary factors) that help moderate the Earth's climate. This will be discussed later.

---

[18]   Do seasons matter? Ask a Floridian living in lovely Miami. She will tell you what she misses most. The denizens of South Florida with their year round balmy weather long for the "Four Seasons" that the people in more Northerly states take for granted. And is there any human equivalent of the ecstasy of an Indian farmer at the approach of the monsoon rains?

CHAPTER 4

THE CONCEPT OF DYNAMIC BALANCE: RUNAWAY HEATING AND COOLING

The earth's overall climate is, and has been for millennia, quite stable; daily, seasonal and year to year variations in the components of the climate such as temperatures (on land, in and over the lakes and oceans and in atmosphere[19]), precipitation and winds, etc., stay within narrow limits. Not only that, but because of Earth's rotation, the tilt of its axis and the resultant seasonality, temperatures between different geographical regions (tropical, sub-tropical, temperate and polar) do not stray as far from each other as they otherwise might. For instance, on the Moon, not only do the daylight temperatures exceed 400 degrees Fahrenheit and the nighttime temperatures approach absolute zero (*minus 350°!*, a swing of some 750 degrees), such changes are extremely

---

[19]   Atmospheric temperatures form a very complex and poorly understood system. As one moves up the lower atmosphere ('troposphere'), the temperatures steadily decrease. This tendency reverses rather abruptly at about 15 Km above sea level (the 'stratosphere'). Several such reversals are noted as one approaches the vicinity of outer space. Even during the most drastic phases of changed global climate over hundreds of millennia, the patterns of temperatures at various levels of the atmosphere do not seem to have changed any at all. This is one more strong argument, advanced in footnote 8, against changes in the rate of insolation caused by varying energy output of Sun.

abrupt and that kind of temperature variation is seen even on the sunlit side between open areas and areas behind shadows of hills and large rocks. Such shocks are, of course, completely incompatible with life.

What is probably not appreciated is that stability of the global climate is not a passive or static phenomenon. Rather, a given state of temperature is the result of the *balance* between forces that tend to raise the temperature and the opposing forces (that would reduce the temperature). This is the concept of the balance of forces. If one of the forces gains an upper hand for a long term, extreme heating or cooling may occur. This is called *runaway* heating or freezing. Of course even the runaway heating or cooling cannot go on forever and a new balance will eventually be established, but at levels that may not be very pleasant. The murderously hot temperatures prevailing over Venus are also now in a state of dynamic balance, although that balance must have been preceded by relatively more equable conditions that got overwhelmed[20] (perhaps!) by warming forces resulting in the runaway heating of its atmosphere.

If the dynamic balance prevailing on our planet is ever lost because the forces of heating or cooling get an upper hand, the new balance would surely be reached again and we might have another ice age or a global climate settling at some very hot state. The resultant temperature may be life destroying. Although such a runaway warming would certainly not reach the levels found on Venus, achieving stability at even a few degrees higher than now would have disastrous results.

---

[20] There is more to this than may appear at first blush. On our planet Earth, a very important mechanism of temperature control involves the so-called carbon cycle. Steady warming (or cooling) impacts the carbon cycle but excessive, runaway heating may not only stop the machine of carbon cycle but may wreck the whole machinery. That is what seems to have happened on Venus. Reversing the runaway warming of Venus (and for that matter, of Earth, should that ever occur) may not be possible without rebuilding all the elements of the Carbon cycle, a very tall order indeed.

Why should we fear such an eventuality? Because both overheating and overcooling are self perpetuating phenomenon, once the 'tipping point' is passed. A little warming sets in motion processes which lead to further warming and so on. Likewise, a little cooling can set the stage for further cooling and yet further cooling until we have a full blown ice age. I will explain later what those forces are and how we might harness them to our advantage.

When we hear so many experts and pundits warn about global warming, many people are skeptical. Simply talking in terms of doomsday does not convince people unless someone explains the forces that come into play to create the catastrophe about which we are warned. And when they inject the role of human agency (excessive production of 'green-house gases' by accelerating the burning of fossil fuels in pursuit of better living standards for fast growing human populations everywhere), ideological and even religious prejudices tend to intervene and close some minds.

You cannot blame people for their skepticism nor fault them for their fatalism about the inevitability of the catastrophe because the same scientists also confidently assert the occurrence of extremes of climate changes in the distant past when *Homo sapiens* was not a significant factor. What is more infuriating to many deniers of impending global warming catastrophe is the talk of a coming ice age that was fashionable not quite a generation ago! And a mini-ice age may indeed precede the much dreaded global warming after all. My good friend Prof. Gordon Johnson of the University of Houston suggested such a scenario: The contention is that if there was to be a repeat of the so-called 'Maunder Minimum,' a seventy-five-year-long spell of cruelly cold climate in Europe (and, perhaps, in North America) from 1640 to 1715, we may forget about curbing our appetite for fossil fuels and revert to unloading carbon dioxide into our atmosphere with reckless abandon. Then, when the cold spell naturally ended, we would discover that the greenhouse gases had, meanwhile, reached irreversibly 'toxic' levels and the global warming was to unfold in earnest. Mankind would then, helplessly, be watching a runaway heating because we may have passed the tipping point while in the grips of a short-lived Little Ice Age! More on the 'Maunder Minimum' and related phenomena later (see Appendix D).

The skeptics ask several questions. What proof do we really have that the energy output of the Sun does not increase or decrease significantly enough to account for the wild gyrations in the global climate? Is it possible that our entire solar system, in its peregrinations in the galactic neighborhood, passes through clouds of matter that would block some of the solar energy from reaching the Earth? And what about oscillating terrestrial magnetic poles or speeding up or slowing down the rate of radioactive fission in Earth's core, a real source of ground heat?

Some of these questions have been partly tackled, others are not even acknowledged by the scientific community. They all need thoughtful reflection and intensive research.

Returning to the subject of balancing opposing forces, we need to explain to the public what compensating mechanisms are triggered when one set of forces appears to be gaining an upper hand and how these compensating mechanisms unleash the opposing forces which balance things out. In fact, these compensating mechanisms (sometimes called 'servo' mechanisms by mechanical engineers) are repeated everywhere in nature, especially in living things. They ensure *homeostasis,*[21] so to speak. Some of these compensating mechanisms are physical-chemical and some are biological. We shall also discuss those forces in detail later on.

---

[21]   Claude Bernard, that great icon of mid-nineteenth century biology and medical science, coined this term to characterize the phenomena that keep every parameter – physical and chemical – of the living beings' internal environment ('milieu interior') within very narrow limits. These include, but are not limited to, the temperature, pH, osmolality and concentrations of various substances.

CHAPTER 5

SECONDARY FACTORS IMPACTING THE GLOBAL CLIMATE

Altitude, topography and shapes, sizes and relative positions of land masses and large bodies of water such as lakes, seas, and oceans all play a role in determining the *regional* and *global* climate. They do so in two ways. Some are immutable, inevitable physical results of their being what they are. The other ways in which they impact the climate and *weather* are dynamic and organic, so to speak, and are mediated through their influence on wind and water currents. Actually, some of the most potent engines of global and regional climatic stability are wind and ocean currents, a fortunate happenstance which does not appear on some other planets such as Mercury or Moon[22]. These *moving*

---

[22] The Moon is not generally considered a planet but a satellite of our Earth. In reality, however, Earth-Moon is a double planet system. This system is completely unique to our solar system (but almost certainly not unique when compared to other solar systems, some of which are bound to have a similar arrangement). We are taught in schools that Moon revolves around Earth; in reality, Earth and Moon revolve around their common barocenter (centers of gravity). As a result, Earth's passage around the Sun does not follow a smooth, nearly circular line but a wavy course.

engines of global climate regulation are lumped together as the *tertiary* factors and will be discussed in a separate chapter.

Before picking up and continuing with the main thread of this discourse, we must digress temporarily to address some of the dramatic but, finally, inconsequential[23] natural phenomenon, even catastrophic phenomena. We take them up here only to dispose of them smartly.

These cataclysms affect *weather* and/or the climate in somewhat irregular, unpredictable and spasmodic ways. These also include asteroid collisions, hits by comets, etc.

Large tectonic upheavals such as volcanoes (rarely earthquakes) that occur during a human life span have dramatic effects on the weather locally or even globally. Yet most probably they have no lasting effect on the global climate unless they produce large-scale and permanent alteration in topography. The nineteenth century eruption of the Indonesian volcano Krakatau (Krakatoa) enveloped the entire planet with ash and soot and blocked out the Sun for several months. As a result, Boston had snowfall on the Fourth of July and there was no summer at all in many parts of the world. The resulting crop failures caused widespread famines and starvation. Yet the planet recovered rather rapidly

---

The reason for this interesting observation is the large size of our Moon compared to Earth. Moon's mass is 1.25 percent of Earth. This fraction is much larger than any satellite of other planets.

If the relative mass were the criterion for down-grading Pluto from its status as a planet, Moon's status should be elevated to that of a planet and not merely a satellite of Earth. But this is really nit-picking. Nature simply ignores the semantic foibles of human astronomers!

[23] By inconsequential I mean only as they impact the global climate and not as they impact the human condition or even long term human history. Major earthquakes and volcanoes have, on innumerable occasions, rudely changed the course of human destiny, ruining some nations while benefiting others, at least indirectly. Volcanic eruption involving the Mediterranean island of Santorini may have been behind the parting of the Red Sea that enabled Moses to escape the pursuing armies of Pharaoh Ramses II out of Egypt!

and completely from that catastrophe. The tectonic upheavals – such as the break-up of the super-continent of Gondwanaland and drifting of the Indian subcontinent against the underbelly of the Asian mainland (part of the Laurasian mega-continent) some two hundred million years ago – are believed to have caused the Himalayas to rise as tall as they are. This radical change in topography has undoubtedly had a very profound impact on the climate of the landmasses on both sides of the Himalayas.

We do not know if there is enough tectonic mischief left deep inside the Earth's interior to cause similar alterations, but, if so, this would stretch out over millions of years and need not worry us. A recent account[24] of the presence of a potential *super volcano* lurking under the Yellowstone National Park in Wyoming, USA and, maybe other places on earth, may cause us to revise our position on this subject, however. Routine earthquakes – they can cause catastrophic loss of life and damage to property – are really not known to have caused many disturbances to alter the climate radically over historical times.

One must wonder if a significantly large asteroid impact is in our future. Many scientists have warned us not to rule out such a monstrous catastrophe in our lifetime. It is generally agreed that a huge asteroid struck our planet some sixty five million years ago, striking the Earth in the region of the Gulf of Mexico but producing a cataclysmic change world-wide. Many life forms were completely extinguished and the dinosaurs may have been the most celebrated victims of that asteroid. The Gulf Stream, the largest of the warm water rivers in the Atlantic Ocean, may have had its genesis in that impact. It continues to shape and dominate the climate of the Eastern U.S. seaboard, specifically Florida and Western Europe.

We do have the means to detect and perhaps deflect or destroy a large asteroid before it makes contact with our planet.

---

[24]   "Will Yellowstone Explode Again?" by Joel Achenbach: National Geographic. August 2009

International organizations are continuously scouting the whole sky, and have detected a few large rocks on the way to rendezvous with the vicinity of the Earth. So far, none is thought to threaten our home planet. A few more years without a significant threat and we should be technologically advanced enough to meet any threats from the sky.[25]

The comets may be a somewhat different matter. In 1903, a hefty comet struck the sparsely populated Tunguska region of Siberia. People did not understand exactly what had happened for many years after the event, but the destruction was frightening enough to excite scientific curiosity. Detailed examination, mostly by Soviet scientists, of the facts surrounding the event eventually clarified the nature of the calamity. Could it happen again? There does not appear to be any comet around that would cause us to worry. Because the comets are illuminated by sunlight and therefore easier to spot at great distances, the statistical odds are that a celestial object would more likely hit the oceans, and we should also be reassured because the comets are made up of water ice. However, if a block of ice weighing several million tons hit, say Manhattan, at the speed of fifteen miles per second, the damage would certainly be apocalyptic and widespread. And even if such a massive comet struck in the middle of a great ocean, it would, at the very least, generate a Tsunami from ten to hundreds of feet high (the great Indian Ocean Tsunami of 2006 was estimated at no more than 30 feet high).

The comets are thought to be the source of water that filled up the low-lying areas of Earth to make up the oceans. In that sense the comets do make the climate, but over very long periods, perhaps several hundred million years. Otherwise, their

---

[25] Really? We may be assuming too much here. Our good luck will be contingent upon our wisdom. We could be united as a species and nothing would be impossible for us to achieve. Or we can succumb to warfare engendered by tribalism (nationalism) and annihilate one another so that when the cosmic catastrophe threatens we may be without means to deploy corrective steps.

immediate impact on global climate can be dismissed as being negligible.

We have largely disposed off the less significant of the natural phenomena as they impact the global climate. We can now examine the more robust parts of the engine of global climate. Some parts of the engine are more or less static and passive; this is discussed here first, leaving for last the most dynamic parts of the motor for examination, in a separate chapter.

So what are the *Secondary Factors* of real significance?

Consider these facts. The Sahara Desert, the largest of the most desolate regions of the Earth, lies at roughly the same latitudes as Southeast Asia and the Indian subcontinent. The southern foothills of the Himalayas receive some of the heaviest (seasonal) rainfalls in the world while at the northern foothills we find the accursed cold deserts of Pamir. The paradox of a mighty river Nile coursing the Nubian-Egyptian desert for some two thousand miles has evoked wonderment for people far beyond the borders of that historic land of fantasy. Not only the Egyptians but the Greeks, Romans and modern day Europeans speculated for ages about the source of so much water in a land of so little rain. The clockwork regularity of the Monsoon rains of the Indian subcontinent and of the trade winds and Horses' winds have served as indispensable building blocks of human civilization, commerce, imagination and poetry. The annual late summer inundations of the Nile – watering the Egyptian desert so faithfully and dependably for millennia that a civilization endured longer than any other – could not have stood for one generation had this promise not been kept unbroken. These climatic facts cannot be explained on the basis of all the other factors we have already discussed, but all of them are the direct results of interaction between mountains and seas[26]. If a

---

[26] And yes, the denizens of Buffalo, New York and Cleveland, Ohio can depend upon the 'Lake Effect' Snow. Large bodies of fresh water also weigh heavily on the climate. It is your choice whether to count those huge snowfalls as a bane or a boon!

sufficiently high (say 5,000 feet or greater) mountain range could be made to appear along the course of the Nile (and to the West of it), the Sahara would be so green and productive that a quarter of it would feed the whole of mankind! That is so because the water-laden winds blowing East and southeast from the Atlantic ocean and Mediterranean sea would be forced to give up their moisture in the form of rains over the entire region west of that range.

Mountains, those outcroppings of land roughly in the midst of the continents, are the lifeblood of nations in more than one way. Their single greatest gift is their ability to wrench out moisture on their windward side just as they assure desertification on their leeward side. Cherapoonji in India and some places in Hawaii claim annual rainfalls of more than 400 inches. These places are extreme examples of the combination of existence in the foothills of high mountains receiving moisture-laden winds from large bodies of water such as the Arabian Sea and Pacific Ocean, respectively. Likewise, the moist southeasterly winds from the Mediterranean blow for thousands of miles over Northeast Africa without shedding their cargo of water until they are intercepted by the Ethiopian highlands where they finally deliver the rains that feed the Blue and White Niles. On the other hand, the lack of mountains of any significant height condemns all but the perimeter lands of Australia to more or less permanent drought, even though all winds blowing over that gigantic island are surely laden with plentiful moisture since they originate over large oceans.

The general elevation of the land, either on the foothills of the continental mountains or just plain elevated plateaus also determines the climate. The higher the elevation, the cooler the clime and vice versa. The consistently hottest places on Earth, such as the Qattara Depression in Libya, Death Valley in California, or the lands around the rim of the Dead Sea in Palestine are peculiarities of geography with one thing in common. All are well below sea level. Conversely, the government of Saudi Arabia, during the summer months, transfers its capital from Riyadh several degrees *closer* to the equator, to *al*-Taif, because, at 6,000 feet above sea level it is measurably cooler and more comfortable than is the sea level Riyadh.

The fact that the general outline of the continents, of the large bodies of water and the mountains hasn't changed except over millions of years, explains the predictable stability of weather and climates as far as they go. Thus the rainforests persist and the deserts have endured during historical times covering hundreds of human life spans. These are the secondary building blocks – all static – that need to be understood as we dissect the phenomena of global climate.

CHAPTER 6

THE TERTIARY BUILDING BLOCKS
OF GLOBAL CLIMATE.

As noted, , the daily local fluctuations in the winds, temperatures and precipitation that we call weather, as well as the overall patterns of these paradigms that account for the climate, are phenomenon that occur in a very narrow band of space above the Earth's surface. They are of such great interest for human beings because we live, work and play in that very narrow band and also share it with almost all the living world. What happens at the extreme depth of oceans – at the most some 35,000 feet, or less than seven miles below sea level – to high mountains – maybe 25,000 feet or five miles above sea level – is all that we (and all animals and plants) care about and this makes life's playground a barely a twelve mile wide 'sphere outside the sphere'. It is composed of the hydrosphere (oceans, lakes and rivers) and part of the atmosphere; and a very small part of the latter at that, namely the *troposphere*. We live by the benevolence of nature as it impacts this biosphere and we may die by its malevolence! No matter what happens here, the universe beyond – both below and above – may barely notice or care.

So when we talk about *global* climate change, we are really talking only about the dislocations of the balancing forces within this twelve mile wide band.

But, of course, this biosphere is very, very important to us humans, animals, and plants. It is also a most delicate piece of machinery. This machinery must perpetually keep on moving, must keep balancing opposing sets of forces and must, by keeping the paradigms of weather and climate within optimal ranges, enable life to go on.

In order to fully appreciate the complexities of the dynamics of weather, we also need to be mindful of the role of two 'cycles', the recurring movements through which water and carbon proceed over and over again. The water cycle is easier to observe and understand. Ocean waters evaporate and saturate the lower reaches of the atmosphere with moisture which is then carried by winds onto land masses where the moisture is disgorged as rain, to the accompaniment of thunder and lightning; this precipitation waters the fields and grasslands and forests and flows into creeks and rivers which directly return most of the rainwater back to oceans and seas, carrying salts from the land into the latter so that their salinity keeps building up. Water captured by plants (and animals) is also eventually returned to the oceans in roundabout ways so that there is no net change in the *'hydrosphere'*. In the process, water does distribute heat from warmer zones to colder belts. We teach pupils even in elementary schools about this cycle because it is so clearly evident and so easy to understand.

Not so with the *'carbon cycle'*; this is far more complex and not intuitively evident. We had to wait till we understood the chemistry of elements and chemical compounds and chemical reactions and, more importantly, the almost magical properties of the green pigment *chlorophyll* and the phenomenon of photosynthesis. But now that we know chemical phenomena in great detail, we understand that carbon, a very pivotal element in life processes, also moves in cycles. Oxide compound of carbon – carbon dioxide – present in the atmosphere is continuously removed from it by the plants – the immense grasslands that once covered the planet – and what is left of them now – and the trees in the forests –through the process of photosynthesis which requires sunlight. Carbon dioxide thus removed gets incorporated into carbohydrates which constitute food for animals. When the animal dies, its tissues decompose and release products that are further broken down by the bacteria into carbon dioxide and other simpler organic and inorganic compounds.

Another venue the atmospheric carbon dioxide takes is when CO_2 dissolves in the oceans. From there, CO_2 moves freely back and forth between water and atmosphere, although some of it can be converted into inorganic carbonates from which CO_2 is not so easily yanked loose. Carbonates get incorporated in sedimentary rocks. Theoretically, this constitutes more or less its permanent removal from the carbon cycle.

In other words, large pools of water that constitute our oceans are absolutely indispensable to the functioning of the critical carbon cycle. They virtually constitute a CO_2 'sink'.

If global warming reached such intensity as to cause the great portions of the oceans to evaporate and then be spun out in space, the carbon cycle is permanently destroyed. This is what is meant by the earlier assertion that excessive global warming can not only stop the machine of climate optimization but wreck the machinery itself.[27] That is what seems to have happened on Venus. We are not in any such danger, during the foreseeable future, however.

I have described the hardware of this machinery, the primary and the secondary pieces, in the previous chapters. The fuel for this machinery comes from the Sun and the moving parts – the water and wind currents styled *'tertiary building blocks'* of climate apparatus – form the subject of this chapter. We noted that the secondary building blocks are generally static. The tertiary building blocks, on the other hand, are dynamic. In fact their effectiveness is entirely dependent upon their dynamic movements.

---

[27]   I lifted this phrase from the venerable 19th century Scottish physiologist J.B.S. Haldane. It was Haldane who enunciated the immortal aphorism: 'Anoxia – lack of oxygen - not only stops the machine (of protoplasm), it wrecks the machinery – (of life)'.

Without the agency of these moving parts, all the other factors that contribute to the evening out of the global climate, namely the spherical shape and rotation of our earth, the yearly sojourn of the planet around the Sun, the optimal length of the day and the year,[28] the inclination of Earth's axis that accounts for the seasons and the disposition, shape and elevation of the continents and mountains etc., would not suffice to moderate the climate of the Earth. Without the oceans and the atmosphere, the tropics would still be hellishly hot; the Polar Regions even colder than they are and the temperature swings between day and night would be of the order of one hundred degrees Celsius (the boiling point of 212 degrees Fahrenheit) rather than 20, 30 or, at most, 40 degrees Fahrenheit as now. Not only that, but the temperature rises during the day and falls during the night would

---

[28]   I did not mention the other unit of time that plays such an important role in our life, namely the month because the length of the month is variable; being not quite an integer fraction of the year, it marches out of step with the year and is not really important because it does not account for any climatic phenomena. We have already 'conquered' the Moon and that has robbed that celestial body of its divine mystique, so to speak. But I don't discount its beauty and the power of its phases to excite human imagination and to inspire poets and song-writers. Moon does cause the ocean tides, a phenomenon which has some short term beneficial effect for humans and one very detrimental long term effect on the Earth. The tidal forces are continuously working to slow the rate of Earth's rotation causing the length of the day to grow. Eventually, after a very long time, the Earth day will be longer, as long as the lunar month and then the solar year. At one point, the same face of Earth will continuously face the Sun, the other half remaining in eternal darkness. Don't worry, though, it will take a couple of hundred million years for that to happen. Moon really has no effect on global climate presently. And the seven day week is another matter. It has no basis whatsoever in nature. Being a device of humans, almost certainly concocted to regulate commerce and to ease labor relations in the ancient Egyptian society, it had to be invested with sacred injunction in order to make it stick. The entire living world and all other forces of nature blissfully ignore what we humans consider a universal and indispensable (also divinely enjoined) unit of time!

be extremely abrupt, almost instantaneous. All of these would make life as we know it impossible.

Let us examine in detail how oceans and atmosphere perform their miraculous assignments of limiting the extremes and of countering abruptness in reaching even those limited extremes.

Oceans, Seas and the large lakes owe their magical powers to the unique properties of water. These unique properties are the melting and boiling points of water, the fact that water expands when frozen, the *latent heat* of melting and boiling and the *specific heat* of water. Other qualities which are important (but warrant a separate and more detailed treatment later) are water's nature as *universal solvent* and an arcane law of nature called *Wilson's Law*.

Melting point, boiling point, and latent heat: Water changes from liquid to solid, i.e., it freezes to become ice at a certain temperature that we humans elect to call zero degree Celsius or Centigrade[29]. In the liquid state, applying one calorie of heat to one ml. (or one gram) of water will cause a one degree (Celsius) rise in its temperature. Likewise, taking away one calorie of heat will cause one ml. of water to cool by one degree Celsius. However, at zero degrees Celsius, something unique happens to water. At that temperature, taking out one calorie from one ml. of water does not cause a further drop in water's temperature by one degree. It takes the removal of 79.8 calories from one ml. of water to freeze it even though the temperature will still register at zero degrees! The linear (but somewhat faster because of the lower specific heat of ice) drop in the temperature of ice will resume once it is frozen. Similarly, applying one calorie of heat to one gram of ice will not cause it to melt and register

---

[29]   There is another scale of temperature measurements which scientists use preferentially. It is denoted as so many degrees K (for Kelvin). One degree Celsius is equal to one degree Kelvin but the melting point of water in Kelvin scale is 273 degrees and 0^0-K (the so-called absolute Zero) is minus 273^0 Celsius. Things can't get colder than absolute zero and close to that level of cooling, many spooky things happen to matter. That is, however, the subject for a different book!

a one degree rise in its temperature. Rather it will take almost 80 calories to melt one gram of ice into water, but that water will still be just as cold, at zero degree. Only the next additional calorie will cause the temperature to rise by one degree. This is called the latent heat of melting or of fusion. Something similar but even more pronounced happens near the boiling point of water. When water is at 100 degrees centigrade, further heating causes it to change from liquid to gas. However, this does not occur with the application of merely one calorie per ml.; rather, it takes a relatively large amount of heat energy – 597.8 calories (583.7 calories under atmospheric pressure of 1000 milibars) – to vaporize one ml. of water. This is called the latent heat of boiling or of vaporization. You will need to remember this when this book discusses the full implications of the latent heats of melting and boiling when it elaborates on the phenomenon of global climate change. For now, I want to remind the reader about the principle under which evaporative coolers work. Evaporation of one gram of water – and water evaporation goes on at much lower temperatures also – consumes almost 600 calories of heat and thus removes it from the ambient air. Evaporation of water, at ambient temperatures, occurs on a significant scale only when the humidity is low; that is why that modality of air conditioning is not very useful when humidity is high (e.g. in maritime areas like New York and New Orleans).

The elementary law of physics requires matter to expand with rising temperature and shrink with cooling. Every known substance obeys this law as does water, most of the time. There is a striking exception to this rule[30]. When water freezes, its lower temperature does not cause it to shrink (and hence become more dense) but just the opposite happens. As a result, frozen water (ice) has about 9% greater volume than liquid water of the same quantity (mass). The result is that ice floats to the top of the

---

[30]   Metallic Plutonium is another exception. (vide "Dark Sun" by Richard Rhoads, ch. 10 pp 192)

liquid water rather than drop to the bottom[31]. This is Archimedes Law in action. What you see of an iceberg above the water is only nine percent of it (the proverbial tip of the iceberg); the 91% is lurking below the surface, to strike the unwary mariner's ill-fated ship if he failed to make safe maneuvers.

There are multiple implications of this seemingly weird phenomenon, some beneficial, others not so good. The obvious danger to shipping of the icebergs floating in the northern and southern oceans is well known to anyone who knows about the tragedy of the Titanic. But the frozen surfaces of the fresh water lakes are the delight of ice-skaters (and danger to those who do not heed the risk of drowning for their failure to be wary of the unsuspected patches of thin ice).

On a larger scale, the oceans and lakes do not freeze from bottom up, the fact that ice floats on waters of oceans insulating the liquid water below from freezing any further. This fact is immensely beneficial for the marine life because it prevents marine mammals from drowning and entombment of all aquatic life in the coffin of ice.

The direct implications of this swelling of ice on the global climate are complex and not easy to understand except that we know it takes a great deal of heat removal to cause freezing of water and ice. And so here is one more force that counters excessive cooling,

Of vastly greater significance for the global climate are the facts of high specific heat and, to a lesser but still significant degree, the latent heat of fusion[32].

---

[31] There is one more phenomenon in nature that teases us. Heavy Water Ice (D_2O) does sink when placed on 'ordinary' water. Heavy water has 2 atoms of heavy hydrogen (deuterium) with the atomic weight of 2. The molecular weight of heavy water is 20 as opposed to ordinary water whose molecular weight is 18. In other words it is about 11 percent heavier than normal water, H_2O, and thus frozen D_2O sinks in liquid H_2O!

[32] Not to be confused with the phenomenon of nuclear fusion.

Since it takes a relatively larger amount of thermal energy to heat (i.e. raise the temperature of) water (in its liquid phase) than most other substances, oceans, seas and large lakes act as virtual heat sinks. Water heats slowly and cools equally slowly. When, at summer solstice (June 21 in Northern hemisphere and December 20 in the Southern), the amount of insolation is the greatest, and the oceans and even relatively small inland fresh water lakes are far from their warmest. Actually the waters of the great oceans reach their highest temperatures almost four months after the beginning of the summer and when the winters are almost upon the lands. Similarly, the ocean swimmers in the northern lands would be much less comfortable in May than they would be in November! Thus the relatively warm seas in winter blunt the bite of cold weather and the relatively cooler waters in summer ease the pain of mid-summers' torpid days. Such a moderating influence is more pronounced nearer the maritime regions than much further inland (the so called *'continental'* climates) but is nevertheless an efficient engine of climatic balance.

This maritime effect is, of course, not by conduction, i.e. heat transfer by contiguity but rather by convection, i.e., through the agency of flowing air and water. The warmer air over the tropical regions moves onto the more temperate regions which are, in turn warmed during the summers; the airflow is reversed in the winter where the cold polar winds cool the more sub-tropical regions. For the most part, vast amounts of heat are transferred by gentle winds. Occasionally tropical warming is excessive. The agency of gentle winds is not enough; there is too much heat that needs to be dissipated in a hurry and we get the hurricanes, cyclones, and typhoons[33] It is somewhat as if a ship is taking

---

[33] These three types of storms really belong to a single species of circular winds whose genesis is in the extraordinarily low atmospheric pressure caused by summer warming of atmosphere. They are all really cyclones which is the generic – and scientifically correct – term, the other names being local nomenclature. The cyclones in the Bay of Bengal, the Atlantic hurricanes, and the typhoons of the Western Pacific are

water and the crew is trying to drain it by using buckets. If the ship gets overwhelmed and the bucket by bucket emptying is not enough, the crew activates the powerful pumps and prevents the ship from sinking. Hurricanes and typhoons maybe awesome and very destructive but so far as nature is concerned they are merely the necessary big measures available to restore the balance. Since most of the tropics are oceans, those monstrous winds also pick up enormous quantities of water which become torrential rains upon landfall.

When people concerned about global warming and its dire consequences cite the probability of greatly increased frequency and destructiveness of cyclones; this point can be driven home more convincingly when the genesis of these storms as heat-dissipating phenomena is clearly understood and explained. Indeed, if the global temperatures increased a mere couple of degrees the cyclones would be so frequent and deadly in so many coastal areas that the properties there might no longer be insurable at a reasonable cost. Destructive category 5 and greater cyclones would only compound myriad other consequences of global heating such as inundation and erosion of coastal lowlands by rising sea levels, and famines from desertification and the destruction of crops.

The physical phenomenon underlying heat distribution through the agency of winds is the phenomenon of convection, one of the three methods by which heat is transferred (the other two being conduction and radiation). We can understand such wind currents better because we are familiar with them to varying degrees, at least by looking on television at the destructive aftermath of the likes of the 2005 hurricane Katrina. Convective

---

one and the same phenomenon. From here on we shall use the generic term cyclones. Tornadoes do not fall in this category of winds designed to achieve heat distribution; however, cyclones do spawn tornadoes. Some of the monster tornadoes which cause billions of dollars worth of property damage and even much loss of life in a few short minutes are not the products of the hurricanes or similar phenomena.

transfer of heat through the agency of winds, however, is not the only convective phenomenon. There are other convective currents that we don't see but which are so large that upsetting them would have consequences that we have only now begun to grapple with. Ocean currents are those phenomena.

The best known and most well studied such current is the Atlantic Gulf Stream which originates in the Caribbean Sea at the equator and meanders roughly north and northeast for over five thousand miles before dissipating near Iceland. So large is the quantity of water that flows as part of that warm current that it exceeds the combined total of all the rivers on land by a factor of one hundred! It has been suggested that the main agency of dissipation of the Gulf Stream is the upwelling of highly saline ocean waters from greater depths to nearer the surface, a sort of giant conveyor belt of salt water. It is further speculated that global warming may turn this regime on its head, aborting the Gulf Stream long before it reaches Western Europe. Western Europe owes its equable climate to the benevolence of the warm Gulf Stream and the latter's diversion could bring about a catastrophic new glacial age overspreading much of the Western portion of Eurasia and perhaps Eastern North America. William H. Calvin[34], the author of this hypothesis, carries the idea much further, suggesting that the Europeans, who are much wealthier and endowed with much greater material wherewithal, may, when faced with a new, particularly brutal, if localized, ice age, forcibly seize the lands to the South, expelling, subjugating or exterminating the inhabitants of those hapless countries![35]

---

[34]   Atlantic Monthly January 1998. William H. Calvin – 'The Great Climate Flip-Flop'

[35]   To his credit, Calvin also proposed alternative, much more civilized, remedies in the form of breaking up some ice bridges to release pent up bodies of water behind the Fjords of Scandinavia to reverse the upset of the conveyor belt. The logic of this approach is murky and not fully developed, probably unworkable. It was my perusal of that article

And although best known and better understood, the Gulf Stream is not the only or even the largest or the longest of the great rivers inside the oceans. There are other currents – some warm and some cold – and their effect on the global climate is probably equally weighty. The Antarctic circumpolar current, flowing from West to East, makes a complete circle between the southern oceans and Antarctica. A cold current, it keeps the relatively warmer waters of the South Atlantic, Southern Indian and South Pacific oceans from heating the Antarctic region. That explains why the lands at the South Pole are so much colder than the North Pole. The total volume of water carried by this current is estimated at about 165 times the total amount of water flowing in all the rivers on the land. If something (like global warming) were to disrupt this current, there would be massive warming of Antarctica from the Southern oceans' waters and we would truly have a catastrophe. The melting of Antarctic ice would cause such a rise in the ocean levels that we might lose as much as ten percent of all land to inundation[36].

We don't know what else global warming will do to the other ocean currents and what, if any, effects their upsetting will have on the ecosystems of the oceans; nor do we know what impact global warming would have on the pattern of winds, rainfall and temperatures. It could be that those changes may partly or fully offset global climate change occasioned by the known forces of global warming. Maybe, maybe not.

At least one other ocean current impacts the climate of an inhabited area of the planet in a discernible way. The so-called California current, a 'cold' current, flows to Hawaii which is cooled by it. Hawaii's climate is sub-tropical although its

---

that moved me to think about alternative ways to counteract global warming, the subject of this book.

[36] Disruption of this Antarctic Circumpolar Current may already have started. The well documented and rapidly developing destruction of the Ross Ice Shelf is probably a by-product of the current's coming demise!

proximity to the equator should make it definitely tropical. It would be nice to investigate the likely effect of global warming on this wind-driven[37] current.

---

[37]   As opposed to the one actuated by upwelling of waters high in salinity, the so-called 'conveyor belt' as exemplified by the Gulf Stream.

CHAPTER 7

SERVO MECHANISMS AND VICIOUS CYCLES.

In chapter two we briefly touched on the phenomenon of compensating mechanisms that cancel out opposing forces of heating and cooling. These mechanisms are called servo mechanisms by engineers and aid the already extant underlying parts that make up the hardware and software of the global climate engine.

Physiologists well understand how the body temperature[38] of human (and all warm blooded animals) remains in an extremely narrow range. These involve shutting off the blood supply to the skin, shivering and assuming heat preserving body attitudes to meet the menace of freezing (in addition to warm clothing, fireplaces and, of course, central heating in the more affluent societies). All these measures work to minimize heat loss and are triggered when the body senses falling ambient temperatures. Similarly, when exposed to excessive heat, the body of *homoeothermic* (i.e. warm-blooded) animals reacts by accelerating blood flow to the skin and sweating in addition to the natural human reflex of discarding unneeded clothing, seeking shade, opening windows to allow a

---

[38]   Not only the body temperature but almost all the paradigms of life – pH, and concentrations of sodium, potassium, glucose and every other physiologic component of body fluids – are kept within narrow ranges by similar phenomena collectively described as homeostasis.

breeze of fresh air, use of fans and, of course air-conditioning, all of which help promote heat dissipation and keep the body from overheating. These forces are triggered automatically – by reflex autonomic action -when the body senses an influx of excess heat (with the exception of human maneuvers dictated by his intelligence).

Servo mechanisms such as the physiological forces described above also exist in physical nature and are triggered when our planet 'senses'[39] forces that might lead to overheating or freezing. The range of temperatures that the globe can withstand is, of course, much wider. The servo mechanism is slow, if eventually massive, and is perhaps not as easily triggered. Most importantly, we don't understand all, or most, of the elements of this servo mechanism. For instance, where are the sensors (*receptors!*) located and how do they function? Partial pressure of CO_2 may be one way, but how about temperature sensing?

One relatively better understood phenomenon is the modulation of the carbon dioxide (CO_2) cycle. CO_2 accounts for only 0.4% of our atmosphere. It is one of several 'green-house' gases but is by no means the largest component. Atmospheric water vapor is the biggest culprit; soot, fine dust, sulphur dioxide, and several other natural components of our atmosphere work together to create the naughty greenhouse phenomenon.

So-called because it mimics what transpires in green-houses (or 'hot-houses', as the horticulturists like to call them), the greenhouse phenomenon is not an unmixed curse although

---

[39]   I may sound metaphysical here but the term is used only because the English language does not provide an alternative word! No, nature does not 'sense' anything the way living organisms sense changes! Most certainly there exist no temperature or pressure 'receptors' in the biosphere that would remotely alert the nuclear furnace in the inner mantle of the earth to speed up or slow down the nuclear reactors in response to temperature extremes in the home of life. Or could there be such a system installed by nature?

perhaps too much of it is (maybe). A greenhouse maintains high temperatures inside even in cold weather because the amount of heat it takes in from sunlight is greater than the amount of heat that is allowed to escape through its transparent walls and roof. The enclosure thus constructed to allow summer vegetables to grow during the dead of winter is warmed by this excess heat.

Clouds have a special role as trappers of heat. We have all learned from our TV meteorologists that the early morning temperatures are much lower when the skies are clear the night before because there are no clouds to prevent radiative escape of heat into space. On the other hand, heavy cloud covers during the day keep the Sun's heat from warming the temperatures as much as they otherwise would. While I was growing up, freezing temperatures were unknown in my hometown. However, I clearly remember a two week spell during which we all shivered day and night. Throughout the period, the days were cloudy and the nights were clear and a few days like that in a row can pile up to create, in the tropical region of the Earth, weather not very different from that of, say, New York.

Greenhouse phenomenon keeps our planet fairly warm. Without it, we would lose more heat to radiation in space than we receive through insolation. If we overplay our hand and curb this benevolent phenomenon too much, we might slip into another ice age.

Be that as it may, the greenhouse 'gases'[40] are now the villain *de jure*. One component in particular, CO_2, receives considerable attention because its excess can be attributed to humans who have been creating large – and exponentially growing quantities

---

[40]   Not all components of the so-called greenhouse gases are actually gases in the strict sense; particulate matter, especially fine dust and, occasionally, volcanic ash, are also members of this criminal gang. When the particles are very, very fine they act as greenhouse gases adding to warming of the atmosphere; when the particles are relatively large, they have the opposite effect of blocking the sunlight and would contribute to cooling!

of it in recent centuries, thanks to our invention of steam and internal combustion engines, and the turbines in the power plants that convert the energy trapped inside the fossil fuels into more manageable electrical energy. It may also be because just as we can accelerate its production, we can also control it (theoretically). And finally, the indictment of CO_2 may have a somewhat religious/ideological tinge to it; isn't hedonistic overconsumption, especially by the affluent westerners, sinful?

The total amount of CO_2 in the atmosphere, although a small percentage of the air we breathe, is still huge, running into trillions of tons. The amount dissolved in the oceans and lakes is several times that and the amount of CO_2 locked up as carbonates and bicarbonates inside the rocks on dry land is even larger. And we dump several hundred million tons of it in the atmosphere every year, far more than ever before. On its face, this should be scary enough. Once again, Mother Nature pulls one of her many tricks out of her hat; the excess CO_2 does not accumulate unchecked and poison our atmosphere. Most of it is removed, blissfully, from the atmosphere by one of two main mechanisms. First, the presence of CO_2 stimulates plant growth. The green pigment chlorophyll goes into high gear and takes up atmospheric CO_2 and converts it into starch in the green, leafy carbohydrate factories of the forests and grasslands. CO_2 thus captured is relatively stable. When starch is consumed by animals, it again become atmospheric CO_2 by respiration – both mechanical and cellular – but much of this starch eventually is incorporated as inorganic minerals, a very stable depository of CO_2.

Another, vastly more efficient mechanism for removal of excess CO_2 is its absorption by the oceans. There is a limit to the amount of CO_2 water can hold in solution but the large bodies of water on this planet are nowhere near reaching their saturation point. Even if they were to do so, many physical, chemical and biological processes would reduce the level of the dissolved CO_2

by incorporating it into inorganic minerals (carbonates) and safely tucking it away in the rocks.

I mentioned Wilson's law in an earlier chapter. That law states that the colder the solvent, the more it can hold of the soluble matters and, conversely, the warmer it gets, and the faster it reaches the saturation levels and starts giving up the solutes either as precipitates (if the solute is solid), or is released as gas in the case of CO_2. If climatic warming were to reach a level where ocean water is saturated with CO_2 and if the other mechanisms for its removal have reached their limits, it would start exhaling large quantities of all gases, including carbon dioxide. That would certainly aggravate the green-house effect and cause further warming of the lands, atmosphere and of waters of the oceans and lakes. That, in turn, would cause large bodies of water to exhale still more CO_2 and then what we have is a vicious cycle (analogous to 'heat stroke' in humans when the compensating mechanisms for cooling the body have been overwhelmed) and hence global warming. Of course hot atmosphere would radiate more heat into space until heat production again equals heat shedding. Undoubtedly the system would once again be brought into balance, but the endpoint may be too uncomfortably hot and incompatible with life. It is this doomsday scenario that provokes the increasingly high decibel warnings from Al Gore and Company.

These movements of carbon – from the atmosphere to organic forms as in vegetations and then through animals and also inorganic pathways through dissolution in oceans and then incorporation into inorganic minerals   is part of the very well understood *carbon cycle*. The carbon cycle is one of the very crucial pillars of climate optimization. It is also extremely resilient and may be more capable of restoring the status quo (despite ungodly insults to it) than we fancy. However, this merciful device can also be utterly destroyed, as seems to have happened on Venus.

There is a flip side to the phenomenon of warming. When the narrow band of hydrosphere and atmosphere in which life flourishes senses too much heat lost, the compensating mechanism of the opposite nature is triggered. And balance is maintained, up to a point. If the oceans were to become significantly colder – for whatever reason – they would remove too many of the greenhouse gases – but mostly CO_2 from the atmosphere because of their increased ability to hold them in solution. Loss of so much of the greenhouse gases would cause too much radiant heat to be lost into space and general cooling would set in. Such cooling would further aggravate the direction of movement of CO_2 and another kind of vicious cycle would set in motion; runaway cooling and eventually a deep freeze that would destroy life as we know it.

In the above discussion, we have concentrated mostly on the role – beneficial as well as malevolent – of greenhouse gases, particularly carbon dioxide. The principal reason for our fixation on greenhouse gases in general and carbon dioxide in particular is that we understand the physics and chemistry involved much better.

It is by no means certain that there are no other regulatory elements to the servo mechanism of climate stability, either with regard to overheating or overcooling. There most certainly are and we need to understand them better. For instance, we don't know what would happen to the ocean currents if we faced climate extremes in either direction. As mentioned earlier, one effect of global warming on the Gulf Stream has been articulated in scholarly articles. That effect, reversing the North Atlantic salt water conveyor belt, would cause the cooling of West Eurasia. You can call that balancing but we would certainly not like that kind of balance![41]

---

[41]   All this only goes to show that humans have very peculiar concerns with regard to what nature's balance is. Nature does not, and doesn't have to, see eye-to-eye with humans over what constitutes beneficial or harmful. Mother nature turns slightly from side to side and we have a deadly tsunami that claims a quarter million human lives. A million species of plants and animals may be made extinct by a single cataclysm such as a volcanic eruption or meteorite impact, but so what? Mother Nature? What Mother?

Another interesting realization, if speculative and almost bordering on science fiction, stems from our understanding of the servo mechanisms and vicious cycles, since the laws underlying all of these cycles are not unique to our Earth and will apply anywhere in the universe. If an unopposed action causing slight cooling of the atmosphere can result in a runaway phenomenon of planetary cooling on Earth, why can't the same thing happen to a planet that is too hot to start with? I am specifically thinking about Venus. If a way could be found to lower the Venusian (Venerean? Cytherean?) temperature, even by a few degrees, couldn't that set in motion a self-sustaining chain reaction (vicious cycle, if you prefer that appellation) that would go on cooling that planet until its climate reaches a new balance at a temperature level that we would find much more genial? It could, if the primary requirement of the presence of a viable, even if very weak, *carbon* cycle can be fulfilled. That speculation will be the subject of a later discussion.

CHAPTER 8

HUMAN CULPABILITY:
REAL OR JUST SO MUCH POPPYCOCK?

Al Gore, the former vice President of the United States and in the eyes of many the legitimate winner of the 2000 U.S. Presidential election, has rightly carved out an honorable place for himself in the history by spearheading the fight against global warming and has been a crusader for saving the planet by educating the public, through the agency of his book, ("An Inconvenient Truth"), the documentary he has produced, and by innumerable lectures and discourses he frequently delivers around the country. In 2007, Al Gore deservedly won, along with Intergovernmental Panel on Climate Chage (IPCC) the Nobel Peace Prize.

Al Gore's interest in the subject of climate change must predate his election defeat because as the Vice President of USA, he is believed to have prodded the National Aeronautics and Space Administration to explore the possibility of a space-based solution to the problem of global warming. That approach, apparently, got nowhere because, shortly after he left that office, the program was terminated by NASA.

The energetic advocacy of Al Gore and many others who inspired him and were, in turn, inspired by him *to do something* about global warming has overshadowed a very legitimate doubt that has been raised about the validity of his assertions about man's role in climate change. That humans have played and continue to play a crucial (and, of course, undesirable) role in accelerating global warming is fast becoming a conventional wisdom. Rapidly accumulating scientific data probably leave very little doubt that the recent measurements of climatic data may not be the usual year to year fluctuations but may herald a true – and very scary – climatic catastrophe about to unfold in our lifetime (at least during the lifetime of our children). All the remaining skeptics are gradually being won over by the proponents of global warming as a real threat. But what if we were leaving out of the equation some significant, as yet unrecognized or ill-understood, piece? Could the crusading of an inspiring and persuasive man turn out to be the walk of the pied piper? We need to keep that possibility in the back of our minds.

While the doubters seem to be losing the argument that a real climate change is under way, there is no complete consensus about the culpability of humans. It is a very real bone of contention and I am not going to put my money on it. There have been humongous fluctuations in Earth's climate many times and lasting thousands of years, in both the directions of cooling and heating. The geological records are replete with evidence for such changes.[42] As recently as the last of those cataclysms, man was nowhere on the radar, so to speak. Only this time around, we have a convenient scapegoat in the form of green-house gases (mostly carbon dioxide) produced by man in ungodly quantities with his burning of fossil fuels (petroleum, natural gas and coal). The rate of fossil fuel burning is only going to accelerate, exponentially, as more and more of humanity seeks to live better and better by consuming more and more energy. The most abundant and therefore the most economical sources of energy, now and for the

---

[42]   Michael Bender, Todd Sowers and Edward Brook: Gases in Ice Cores: Proceedings of the National Academy of Sciences of the United States of America, August 5, 1997 Vol. 94, No. 16. (Besides this very good reference, there are a large number of other, equally credible, scientific articles, too numerous to catalogue in this work).

foreseeable future, remain hydrocarbons derived from mineral sources. For practical and political reasons, renewable energy sources and nuclear energy must remain secondary to fossil fuels.

The recent trend toward the use of biofuels, to augment the fossil fuels, appears sensible because it merely returns to the atmosphere the same carbon dioxide that was 'harvested' out of the atmosphere by the agency of photosynthesis. In effect it is an indirect use of solar energy and there is no net addition to the amount of greenhouse gases. But the same wise solution, albeit a partial one, has also caused worldwide food shortages, runaway inflation in the cost of food in some poorer countries and even food riots. It may have been responsible for accelerated deforestation of the Amazon basin in Brazil where adventurous farmers have started growing for palm oil! Talk about the law of unintended consequences!

A more hopeful note was recently sounded by National Public Radio when, on a July 15, 2009 segment of "All Things Considered," Robert Siegel's interview with J. Craig Venter unveiled the $600 million collaboration between his company Synthetic Genomics and Exxon-Mobile with plans for growing algae as a source of biofuels. This approach reasonably and realistically mitigates the pitfalls of pursuing ethanol from grains and other food sources as a substitute for fossil fuels.

The climate crusaders assert that unless we institute drastic measure to cut CO_2 emissions through education and the carrot of financial incentives to polluters (but mostly through the stick of government regulation, such as the Kyoto Protocol), it may soon be too late. In their opinion, the servo mechanisms that unleash the opposing, self-correcting forces have already been overwhelmed and a vicious cycle of warming contributing to still greater warming, as described in the previous chapter, is now threatening to condemn our planet to a runaway warming – a planetary heat stroke, if you will – that will soon make the planet uninhabitable.

I would say to these well meaning folks, please, hold your horses. We don't quite understand the corrective forces well enough to write them off as already overwhelmed. We also don't have a real feel for the planet's capacity to heal, to overcome the excess 'burden' of greenhouse gases by mechanism yet ill-understood. Nor do we clearly grasp what burden is insurmountable by the processes that we do understand. For instance, high CO_2 levels do stimulate plant growth[43] and additional biomass can certainly take up much of the excess CO_2 through photosynthesis. Also, the oceans have enormous capacity to dissolve excess CO_2 and this capacity increases as the amount of CO_2 in the atmosphere increases. These facts alone, plus the unknown other compensatory responses involving ocean currents, would make the prophets of doom premature in their gloom.

Having said that, I cannot advocate an absolute complacency, either. What if the 'prophets of doom' are right? Can we risk bypassing this narrow window of opportunity to *do something* for our planet's health? Is it not better to err on the side of caution?

The measures and goals implicit (and explicit) in the Kyoto Protocol sound sensible, but I have serious doubts about their feasibility. If 15 percent of the affluent segment of the humanity which lives well by burning fossil fuels has brought us to this sorry pass, where would we be if and when the multitudes in the underdeveloped world begin to approach the level of affluence enjoyed by the developed world? Short of re-imposing colonial rules, how can they be dissuaded or coerced into holding back on their aspirations for a better life (two car garages, a plethora of superhighways, central air-conditioning and heating, and plenty of food to waste)?

---

[43] But then we have another conundrum here. The great forests of the Amazon basin and other regions of the earth are being cut down so fast (in millions of acres every decade) that their excess luxuriance in consequence of elevated CO_2 levels may simply not suffice and the net capture of that compound may be zero or negative.

So we can all agree that something must be done and also that, perhaps, the Kyoto approach is a non-starter (much as I hate to agree with George W. Bush). Is there any other way? Have you ever been to the beach? Have you seen those gaudily colored large umbrellas that protect the sunbathers from overheating and sunburns? Does that tell you something?

And have you ever experienced a total eclipse of the Sun? Or at least read some accounts of it? In the chapters that follow, I will make a case for preventing runaway global warming by blocking a relatively minute portion of insolation incident upon our planet. I will also explore with you how the phenomenon of 'vicious cycle' can be harnessed to serve not only our planet but also, at some future date, to change the climate of other planets (I have no use for science fiction, honest!)

CHAPTER 9

OF TRANSITS, OCCULTATIONS AND ECLIPSES.

Such a jump from a discourse on global climate to matters celestial may appear weird at first, but I need to explain these before I can tie the two subjects together and advance my proposition that really forms the subject of this book. In fact, you really don't need to read the first eight chapters to appreciate the possibilities that I am proposing.

When a celestial body, in its heavenly sojourn, passes between a second and a third celestial body, an observer on the second body will notice that his or her view of the third body is partially blocked. The planet Mercury often courses across the solar disc and this so-called *transit* excites the astronomers because it is such an exhilarating (for them) sight. Astronomical publications announce coming instances of such transits with much fanfare and amateur star gazers organize lively 'star parties' to observe the phenomena.

Likewise, when a celestial body, specifically a planet in our solar system, passes between us (i.e., our Earth) and a distant star, that star is momentarily hidden from our view. This phenomenon is called occultation. No, I am not trying to sell witchcraft or black magic. Occultation is a technical, scientific term seriously used by serious astronomers! They find this to be an *interesting* phenomenon. For some scholarly observers, it might even yield groundbreaking information such as the discovery of an unsuspected planet or a companion star. But I am getting too far afield.

A more common event of astronomical interest is the eclipse; I am specifically referring to the solar eclipse. Partial or total eclipse of the sun is, in reality, the *'transit'* of our moon across the face of the Sun. It can also be likened to its *occultation*.

But unlike transit or occultation, solar eclipse is far more spectacular. When the eclipse is total, meaning that the lunar disc completely covers the solar disc, the effect is said to be dramatic and eerie. Total eclipse of the sun, although infrequent for any given location, is not all that rare. Only it is visible over very small areas of Earth and it lasts for a relatively short time (totality lasts a few minutes). But during those few minutes, the daylight seems to vanish and darkness descends upon the scene when only moments earlier it was full daylight. Myriad stars and planets become visible in the sky in the middle of the day. Atmospheric temperature drops several degrees in short order. Birds return to their perches and fall asleep. Nocturnal creatures spring into action and others go to sleep. And when the crescent-like outline of the Sun emerges from behind the Moon and it is daytime again, roosters actually crow, the animals and birds wake up and resume their activities and the nocturnal ones again retire. So dramatic is the scene that all cultures across the globe, over millennia, have ascribed a divine – sometimes malevolent – hand in the happening and have concocted ways to assuage the gods!

The reason this spectacular celestial event is possible is that the *apparent* sizes of the discs of the Sun and the Moon, as seen from the Earth, are very nearly the same. They both subtend an angle of about half a degree. Now we know that the Sun's diameter is immensely bigger than that of the Moon, and the latter is also a whole lot closer to us than is the former. At its perigee (when closest to Earth), the angle the Moon's disc subtends (about half a degree) is actually slightly bigger than the angle the Sun's disc subtends. Hence it is possible for it to completely obscure (*occult*) the latter. We than have a truly '*total*' eclipse and you can stare directly at the Sun without risking damage to your eyes! Conversely, at its apogee (when farthest from the Earth) the angular size of the Moon's disc is slightly smaller than that of the Sun. When the centers of Sun, Moon and Earth are perfectly aligned we do not see a *total* eclipse. Rather, what we get is an '*annular*' eclipse. A skinny ring of the Sun's disc surrounds that of the Moon. (You still can't look directly at the Sun during an annular eclipse. What is still visible is enough to cause severe retinal damage). None of the marvelous and astonishing things we just described with totality happen in the annular (or partial) eclipse, except for the breathtaking spectacle – better viewed on YouTube or your evening TV newscast –of the eclipse itself.

Looked at another way, a solar eclipse occurs when the moon's shadow falls on earth. A conical inner shadow ('*umbra*'), which is much smaller in diameter than the diameter of the moon, gives us the total eclipse. Outside the *umbra* is the much larger '*penumbra*'. All those areas of the Earth which only fall under the *penumbra* see only a partial eclipse. Both *umbra* and *penumbra* sweep across a narrow swath of the sunlit side of Earth rather rapidly, moving at about 1,600 miles per hour. Because The Earth's surface is mostly covered with seas and oceans, most of the time solar eclipses are visible only over limited areas of the oceans and sometimes the islands and only infrequently on parts of the continental mainland.

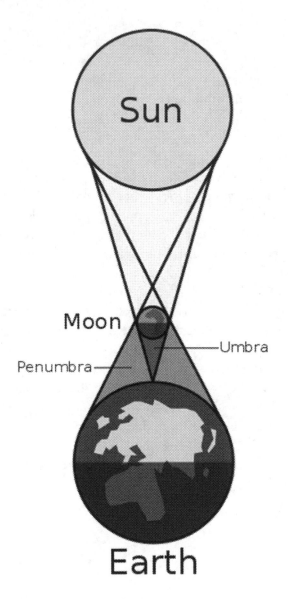

Figure 1   Reproduced from Wikipedia
(No permission required)

The diagram of the phenomenon of solar eclipse depicted in Figure 1 is obviously not to scale. It explains the concepts of *umbra* and *penumbra*. The *umbra* is often no more than 200 miles wide, only about 10 percent of the Moon's diameter, but the *penumbra* can cover a quarter of the whole Earth.

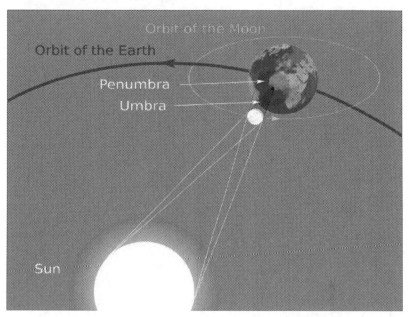

Figure 2   Reproduced from Wikipedia
(No permission required)

Figure 2, above (also not to scale), depicts the phenomenon of solar eclipse in somewhat more realistic way. The reason for the phenomenon of *Umbra* and *Penumbra* is that the aspect of the Sun is so much larger than that of the moon that former's rays do not fall upon the latter as parallel beam. If moon were any smaller or further from the earth, the umbra would not even reach the surface of the earth and we would never see a total eclipse.

When the centers of the Sun, Moon and Earth are disposed over a straight line, they are said to be in conjunction. The corollary of this fact is that a solar eclipse can only occur at New Moon. A lunar eclipse follows solar eclipse, especially a total solar eclipse, half a lunar month later, at Full Moon. A much larger area of Earth is then treated to the lunar eclipse than the solar eclipse.

There are other, more esoteric, scientific aspects to solar eclipses which are often intellectual grand feasts to professional astronomers, but the discussion of those aspects is not germane to the subject at hand.

The point of this discourse over the solar eclipse is simply this. The phenomenon demonstrates, in a dramatic fashion, the role of *insolation* in the genesis of weather. And climate is the sum total of weather. When a beachgoer places an umbrella above her spot on the sand, she is, in effect creating a miniature eclipse. If the umbrella is rotated from East to West to synchronize with the movement of the Sun, you have a sustained eclipse. Blocking the rays of sun, even partially, reduces the amount of heat reaching us. This is too elementary to merit any further elaboration.

If an umbrella of sorts is placed in space between Earth and Sun in such an orbit as to maintain itself *continuously* at a spot over the line running from the center of the Sun to the center of the Earth you would have a permanent solar eclipse. This trajectory can be called '*cum*-sole' or '*Heliosynchronous*' orbit. Even if the *umbra* of that eclipse is very small, say fifty miles or so in diameter, you would have a discernible cut in the amount of solar radiation reaching our planet. If the umbra sweeps daily around the Earth in the tropical regions, between 23.5° north and 23.5° south, the discount would be considerable. The bottom line is that the *occultation* would be continuous, day after day, year after year.

As I indicated at the outset, what I am proposing is modifying one of the five prime factors that underpin the genesis of Earth's climate; specifically the magnitude of insolation – something that appears, at first blush, to be fantastic and bordering on science-fiction. But science-fiction it definitely is not. In the following chapter I will try to make the case for the practical application of this concept in greater detail.

CHAPTER 10

A SUN BARRIER IN A 'HELIOSYNCHRONOUS' ORBIT.

As I explained in the previous chapter, placing a barrier that would maintain its position *continuously* between Sun and Earth could cut the amount of solar radiation reaching the earth. Such reduction, if sustained, could neutralize the alleged warming trend – regardless of whether caused or aggravated by human activities or not – that is threatening our home planet, specifically the biosphere part of it.

Once we accept that thesis, we need to address a number of questions that would immediately arise and try to search for other pesky questions that have not yet surfaced and will definitely become apparent and might plague us as we get along with the project.

Is there a 'natural' version of Heliosynchronous orbit consistent with what we know about celestial mechanics?

One issue that will immediately become apparent is that theoretically the only trajectory that even vaguely resembles the proposed Heliosynchronous orbit is the so called L1 (after Joseph-Louis Lagrange) orbit that maintains a *reasonably* stable locus between a primary and its satellite.[44] And I said 'reasonably'

---

[44]   I will be using the term 'satellite' in a loose sense. We are used to thinking of satellites that orbit planets which in turn orbit their star. In the sense I am proposing, planets are satellites of the star and moons

on purpose, because even L1 orbit, although faithfully Heliosynchronous, will at some time degenerate because it is dynamically unstable. The satellite placed in this orbit will eventually find a stable locus in L4 or L5 positions.

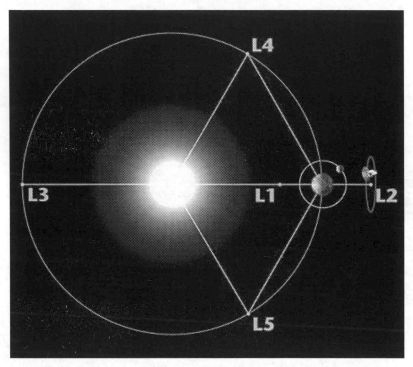

Figure 3: Reproduced from "the Lagrange Points" by Neil J. Cornish of NASA Wilkinson Microwave Anisotropy Probe Laboratories. No permission needed. Reproduction does not imply NASA endorsement of ideas expressed herein.

---

are satellites of planets and there is no reason to doubt that moons could have their own satellite system. Let us not be extreme purists and quibble about specific terms as long as the meaning is made clear.

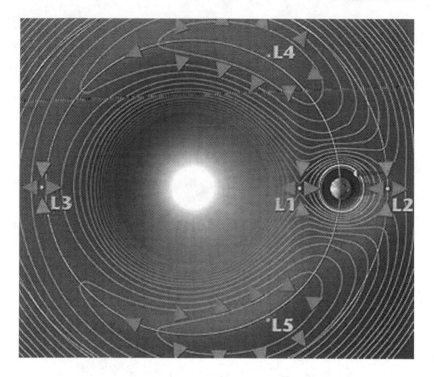

Figure 4
A contour plot of the effective potential of a two-body system (the Sun and Earth here) as viewed from the rotating frame of reference in which Sun and Earth remain stationary. Objects revolving with the same orbital period as the Earth will begin to move according to the arrows indicating the slopes around the five Lagrange points — downhill toward or away from them, but at the points themselves these forces are balanced.
Reproduced from "the Lagrange Points" by Neil J. Cornish of NASA Wilkinson Microwave Anisotropy Probe Laboratories. No permission needed. Reproduction does not imply NASA endorsement of ideas expressed herein.

Figure 3 and 4 depict the various Lagrangian loci, from L1 through L5 in a diagrammatic and in more fanciful ways respectively.

But even if the locus L1 were inherently stable, that is, able to be maintained inertially, without propulsion, it would not serve our purpose. The reason for that is because the L1 locus is way too far, about a million miles, or roughly four times as far as Moon! At that distance, the size of the proposed barrier would have to be humongous, at least 333 miles in diameter, in order to intercept less than one per cent of solar radiation. And because of its inherent instability, it would require powered steering to keep it in the correct locus. The expense, not to mention the technical challenges of such an undertaking, squarely places the concept in the realm of science fiction.

We must discard L1 locus barrier as impractical for now and for many years to come.

A more technically detailed, mathematical depiction of forces involved is further provided by NASA, probably useful only for professionals. However, we do have the ability, wherewithal and technical know-how to place a reasonably sized barrier in a much tighter Heliosynchronous orbit where the object can be forcibly maintained in the desired locus by powered propulsion. The energy required would be quite small, relatively speaking, and could be derived from Sun. Without a doubt, placement in a tighter orbit, much closer to the Earth than our Moon, would have to take into account the lunar gravity; but if we are not depending upon inertial stability anyhow, who cares?

The second question to be addressed concerns the size, material and design. The size would still have to be quite large but only in two dimensions. The main barrier can be exceedingly thin, perhaps on the order of a few meters. The area of the barrier may add up to several hundred square kilometers. Without an adequately large size, the umbra of the barrier may not even reach earth's surface. This would entail construction in space by adding a series of modules joined together in much the same way we are adding modules to the international space station, only much simpler and cheaper. The only substantial part would be the electronics and propulsion system, but the latter would not have to have fuel replenished from time to time. The living quarters for the maintenance crew could be kept in the International Space Station (ISS). Use of robots could minimize need for a large human crew.

As for the materials, there should be no great obstacles. A wide choice of materials for the construction should be available. Some examples could be aluminum, Mylar, other plastics, or a combination of these. Jerome Pearson and associates[3] suggested a wide variety of types and sources for constructing barriers; these were in response to a different concept but remain eminently suitable for our project. The materials can be derived from terrestrial, lunar, or asteroidal sources.

The design of the barrier should allow for reducing the amount of radiation blocked should there be unforeseen effects such as excessive cooling. A louvered design is one possibility. Merely reorienting the barrier to make it face the sun edgewise might prove to be the easiest solution.

Once placed in orbit and engineered for appropriate propulsion system, the device should require very little care and maintenance. But read the next chapter for caveats.

What may we expect from a properly placed and functioning Heliosynchronous barrier? First off, we would have a measurable decrease in the amount of solar radiation incident upon Earth; a sustained cut in insolation would inevitably reduce the atmospheric temperature immediately and cause a cooling of oceans more gradually but with equal certainty. The capacity of the oceans to hold CO_2 in solution would increase, resulting in the migration of huge amounts of CO_2 to move into solution, thus reducing their greenhouse effect. A further atmospheric cooling and an ongoing trend in that direction will immediately remove the threat of global warming and the nasty upheavals that are foreseen to follow in its train. At some point in the future, we may have to worry about excessive cooling but we have the means to 'turn off' our friendly little barrier by reorienting it.

One should contemplate the comparative merits of this approach with those of other measures being promoted to counter global warming, such as those outlined in the Kyoto Protocol.[45] At the heart of the protocol is international cooperation to reduce CO_2 emissions by all nations, chiefly by controlling the use of fossil fuels. One device proposed is that of CO_2 vouchers. This last suggestion appears to me to be mere gimmick – almost Quixotic in its optimism – and cannot have a long term positive effect on the goal of CO_2 reduction. The clamor for higher living standards among the populations of the underdeveloped countries (and those living on the fringe of poverty in the developed nations, too) is just too potent to squelch and they form more than 80% of humanity. The problems associated with the use of bio-fuels have already been mentioned. It may be possible to substitute non-food sources of bio-fuels such as harnessing cellulose digesting bacteria and thus converting it into usable fuel, but the small quantity of available waste cellulose may be the limiting factor. Wind, solar, and nuclear power are good alternatives and should definitely be pursued, but they cannot address the problem as big as that posed by the explosion of rising expectations.

If the governmental leaders can be persuaded to make the modest investment required for this project, the downside risk associated with any project to block the sun's heat appears quite small.

In a later chapter we will turn to the last issue which excites me a great deal. It takes the possibilities of solar barrier beyond our home planet. But for now we need to examine the issue of *'Heliosynchronous'* Orbit.

---

[45]   The 1996 Kyoto Protocol (discussed in Appendix D) signed by all the nation states and also ratified by almost all of the signatories and the new covenant to replace it when it expires in 2012.

CHAPTER 11

HELIOSYNCHRONOUS? ORBIT?

The title of this chapter challenges the very concept of 'heliosynchronicity' of any trajectory. And when this issue is examined more closely, the word 'orbit' itself will become suspect. I am addressing this issue here to ward of any criticism that may unjustly and prematurely discredit the underlying concept of a 'permanent' solar barrier placed in space.

The term heliosynchronous is a composite of two Greek nouns (*Helios*, meaning Sun and *Chronos* meaning time) and a Greek conjunctive *(Syn* meaning with or in step with[46]). The space barrier designed to continuously block a portion of solar radiation from reaching (i.e. being incident upon, if you want technical terminology) the earth would have to be, from the point of view of us living on the earth, continuously in step with movement of the sun. In other words, it can be described as *'cum sole'*. The obvious inspiration for this term is *'geosynchronous'*, as in geosynchronous communications satellites which already circle our planets in very large numbers. Without going into the mathematics or mechanics of these orbits, it is sufficient to say that a satellite placed in orbit around Earth (or any celestial

---

[46]   Compare this with the Latin, 'sin' which has the exact opposite meaning! Spooky, isn't it?

body) completes one orbit around it in a specific period of time. This amount of time (technically called orbital period or, simply, the period) depends upon the distance between the parent body and the satellite. The farther away the satellite is placed, the longer is its period. In the case othe earth, the period cannot be shorter than 90 minutes because, if the satellite were placed any closer, it would fall back on the Earth after completing only a few revolutions. The trajectory of such a 'doomed' satellite is described as a '*cycloid*'.

You can place the satellite as far above the surface of the Earth as you care to. The length of its period will grow accordingly. At a point when the period of the satellite matches the length of Earth day (approximately 24 hours), the satellite is said to be in a geosynchronous orbit. It then stays above the earth over the same geographic location because it revolves around the Earth at the same rate as a given point on Earth rotates around the planet's axis. In other words, a geosynchronous satellite is in step with the rotation of the earth This is the basis for a telecommunications satellite that bring us the television shows captured by the ubiquitous satellite dishes that now dot the landscapes of practically every locality in every country. The distance at which a satellite becomes geosynchronous is about 23,000 miles. The navigation satellites that underpin the now ubiquitous Global Positioning System are all geosynchronous satellites (although they don't have to be).

If you adhere to linguistically pure definition of '*heliosynchronicity*', then a body would be a) implicitly circling, primarily, the Sun and b) the orbital period of that body would be the same as the time it takes Sun to rotate on *its* own axis (approximately 25 to 40 days! Sun's equatorial region moves much faster than its laggard poles). No go, obviously. The first objection, of course, is inconsequential and more in the nature of nit-picking since anything that orbits the Earth also revolved around the Sun. But the second objection is insurmountable, at least semantically. Not only that, the Sun's period of rotation around its own axis is by no means simple. Not only do the equatorial regions of the Sun spin faster than the Polar Regions, various layers of Sun rotate at disparate velocities (which fact explains the monstrously intense magnetic field of that star)!

I am therefore proposing that we drop the term *'Heliosynchronous'* altogether and adopt the term *'al Battani'* trajectory. The problem with the term orbit, which we will discuss shortly, is insurmountable but the designation *'Al-Battani'* is eminently well deserved because it honors the remarkable life and work of the ninth century giant of Samarra, Iraq. Appendix A narrates many details about the life and contributions of that remarkable Arab astronomer of (Europe's) "Dark Ages." It is also proposed that the barrier placed in that kind of trajectory be called the *Al-Battani Shield.*

As I said, the term 'orbit' cannot fairly be applied to any trajectory not dependent solely on *inertia* but maintained by continuous application of external force. A barrier placed, say, 40,000 to 200,000 miles above the Earth – but gravitationally linked to the sun –if left to inertial forces, will not maintain its position between the Earth and the Sun and must continuously be nudged in place by firing of especially weak rockets. The *sidereal year* of this kind of object (literally a planet rather than a satellite!) will, of necessity, be shorter than that of the earth. It will therefore tend continuously to race past the earth and would have to be 'slowed down' with retro-firing thrusters without falling out of the orbit. These rockets could be ion rockets and able to derive the energy needed from the Sun but I doubt if they could ever be on auto-pilot. With these caveats fully understood, there is no inherent scientific reason why it could not be done.

Al-Battani orbit can only be applied to an object orbiting the Sun and not the Earth. Let us visit some basics of our solar system.

There are eight major planets orbiting our Sun. (OK, nine if you still insist Pluto is a planet). The time it takes each one to complete one full orbit (so-called orbital period or, simply, the period equivalent to the sidereal year of that planet) depends directly on the planet's distance from the Sun; the closer it is to the Sun the shorter its period and conversely those orbiting the Sun very far out take much longer to complete their sojourn. The table below shows the periods of the planets Mercury through Neptune. The second column denotes the planet's mean distance from the Sun in astronomical units. An astronomical unit, as mentioned earlier, is the *mean* distance between Earth and Sun (approximately 93 million miles).

Planet	Distance from Sun	Orbital Period (in Earth days or years)
Mercury	0.4 AU	88 days
Venus	0.7 AU	225 days
Earth	1 AU	365.24 days
Mars	1.5 AU	687 days
Ceres[47]	2.77 AU	4.6 years
Jupiter	5.2 AU	11.86 years
Saturn	9.5 AU	29.66 years
Uranus	19.6 AU	84.3 years
Neptune	30 AU	164.79 years

It is evident that a body in *Al-Battani* orbit must have the same orbital period as that of Earth and its distance from the Sun must be very slightly less than 1AU. This is not possible inertially. Even if we placed the barrier at, say .9999 AU from the Sun (approx. 9,300 miles from earth but that much closer to the sun),

---

[47] Some Plutophiles will cry foul here. Drop Pluto and include Ceres – a mere asteroid? What heresy! But according to the professionals, Ceres is more planet-like than Pluto (really a runaway satellite of Neptune.) And it is quite massive, capable of being seen using really small telescopes. It is a remnant of a classical, good-sized planet that used to orbit the sun between Mars and Jupiter but had the misfortune of being shattered to pieces by some yet undetermined agency hundreds of millions or a few billions of years ago.

its period would be slightly less than one sidereal year. The result would be the loss of *cum sole* location within a few days.

What that means is that the period differential between Earth and *Al-Battani* shield would necessitate constant correction to be accomplished, perhaps, with on-board ion thrusters.

Since the *Al-Battani* shield would be in orbit around the Sun and not in Earth orbit, the impact of the Earth's gravity would have to be figured in calculating the force and timing of corrective thrusters.

But these are not the only mechanical considerations. We need to take in account the *eccentricity* of Earth's orbit. Since Earth's orbit is not perfectly circular, the distance between Earth and Sun can vary by as much as three million miles. Unless the eccentricity of the shield's trajectory were exactly the same and its phase also exactly congruent with that of the earth, the fluctuation in its distance would be great enough to render the barrier's effectiveness marginal at times.

And lastly we need to take into account our Moon. At the distance proposed (.9999 AU), and with the limited size, the barrier would not be too vulnerable to perturbation engendered by the gravity of Moon. However, the paths of Moon and of our barrier must inevitably cross at least every few months. Without provisions, first avoiding collision and then restoring the barrier in its desired trajectory would require a great deal of computations and constant adjustments. These are some of the most obvious obstacles to achieving successful placement of a barrier in the proposed trajectory.

How much of the Sun's radiation need we block? And what disasters might strike our mother-planet from such reckless tampering with Mother Nature?

Very detailed computer models with enormous amounts of data and innumerable scenarios fed into supercomputers alone can give precise answers to these questions. We need to remember that with a 'permanent' occultation of the Sun by a man-made barrier, the umbra of the occulting (or eclipsing) body would still be extremely small and it would sweep the planet very rapidly. The umbra would stay over the tropics and cover different spots on Earth on different days.

Common sense tells us that the amount of insolation needed to be blocked would be very very small, perhaps a small fraction of one percent. We are not seeking runaway cooling of the planet but only trying to avert its runaway warming; in other words, we aim to preserve the *status quo*. We also accept the possibility that at some point we may have to cease and desist from this kind of 'mischief'. All these are a given. Wisdom and humility are not incompatible with bold vision.

There is no doubt about the scientific merit and technological possibilities. If we can summon available resources and requisite consensus in time, a practical and affordable solution to the problem of global climate change may be at hand. Additionally, I believe that this solution may be more politically acceptable, economically affordable, and technologically achievable than some alternate solutions that are currently under serious discussion, such as issuing 'carbon vouchers' to major polluters and allowing them to trade such vouchers among themselves. Implementation of the search for renewable energy sources which do not depend upon the burning of fossil fuels are, of course, an exception from the above assertion and must be vigorously pursued.

One more thought. The economics of undertaking construction of an *al Battani Shield* must appear such as to make any such project impossible under present circumstances. Global recession and, especially, near bankruptcy of the wealthiest nation in the world is already threatening severe curtailment in the manned space program of the United States. By some estimation, because of the lack of funds to replace the soon-to-be-retired space shuttle fleet, it is feared that manning and maintenance of the International Space Station may itself have to be put in abeyance for as long as five years.

However, if we pool all the resources currently dedicated to the military endeavors of all the nations[48] (1.2 trillion dollars in 2008, at current estimate) and decide to take on the challenge of global warming as one united species and to safeguard the future of mankind, nothing may be impossible. Would NASA welcome a challenge bigger than the challenge of landing a man on the Moon and bringing him back, especially if it did not have to waste its budget on developing and deploying space weaponry and, more especially, if it could draw on the resources of all nations toward a unified goal such as outlined here? I bet it would.

---

[48]   Am I whistling in the wind? Renunciation – or even curtailment- of military machines by the nation states may be even more far-fetched that asking the less affluent to defer, let alone forego, their dreams of better living standards.

CHAPTER 12.

HOW ABOUT THE OTHER WORLDS?

We now come to an area which is really speculative and intellectually challenging. There is neither urgency nor any long-term utility to the ideas developed here, just mental exercises for the curious. And so, if you want to relegate the following to the arena of science-fiction, please be my guest.

A term *terra-forming* has recently entered the lexicon of science-fiction 'crackpots' (they would prefer the epithet *futurists)*. The term suggests engineering endeavors to cause the conditions on other worlds to resemble those on the earth to the point that long-term or permanent colonization by humans may become a reality. Most of the discussions on web sites such as *space. com* have centered on the planet Mars as the target of 'terra-forming' and I really have no problem with this. There are several attractive elements about the red planet. It is reasonably large, close enough to the Sun to receive enough heat for maintenance of life. It has an atmosphere (albeit very thin and poor in oxygen) and may have sufficient water to satisfy the needs of a few thousand humans – under strict rationing and recirculation regimes and for a vestigial form of agriculture. Most importantly, the lengths of its day and year are very nearly identical to the Earth's. Besides, its axis of rotation is tilted almost to the same degree as our planet – although not pointing at our pole star -

so that there is an earth-like seasonality to its year. There might be enough native materials available for us to build just about anything we might care to build there. Of course, Mars does not have the large beautiful Moon like ours; its two moons are much smaller, are only faintly visible at times, and they go around the planet at a much more hurried pace and will not excite poets and artists to romance. The unique view of the Earth/Moon double planet system from Mars, with its inevitable moon-like phases, may partially compensate for the rudeness of the loss. But there are really no other insurmountable obstacles to colonizing that planet.

My interest in planetary *'engineering'* has been directed to another planet, namely Venus. That planet has sometimes been called Earth's sister planet because of its size, which is nearly identical to that of our fair mother planet. Aside from its size and reasonably similar distance from the Sun (one third closer), this sister of Mother Earth has very little in common with the Mama. Human beings visiting this world would find her to be a very hostile aunt indeed.

The single most important element of the fierceness of Venus is the temperature of her toxic atmosphere which is about a thousand degrees hot and known to be 95 percent CO_2, a poster child for global warming run amok. There is a challenge there but also an opportunity for humans. Very high concentration of CO_2 and the furnace-like temperatures of the Venusian atmosphere are two aspects of the same problem. If a process can be set underway to start cooling the atmosphere, reduction in the amount of CO_2 would follow (and vice versa).

We saw in an earlier chapter, how we can reduce the atmospheric temperature on our planet coupled with removal of carbon dioxide by blocking insolation with an artificial barrier in a quasi-helio-synchronous (Al-Battani) trajectory. If the concept is solid with regard to one planet, it should be comparable for another.

There is the pesky problem of the magnitude of the job; on Earth, we are talking about reducing both the CO_2 concentration and the temperature by minuscule amounts really. In the case of Venus, the task is herculean. If it would take a couple of years of discounted insolation to arrest global warming on Earth, it would take centuries on Venus. But are we in a hurry to 'fix' Venus? Not really.

But even if we did cool Venus down to earth-like temperatures, she would still not be a very friendly aunt (or second mother). It has no seasons and so no enhanced heat distribution; its day is too long – longer than its year. Can you wait 2,784 hours to watch the sunrise in the West?

Venus rotates very slowly – once every 243 Earth days – on its axis and it rotates backwards. It goes around the Sun once every 225 days. So the Venusian day is longer than its year but because the path of its revolution is pro-grade (in the same direction, anticlockwise if observed from the North, as all other planets and most satellites) and its rotation around its axis is retrograde, the Sunrise and sunset occur every 116 Earth days. Venus has no Moon. Both these weird facts probably are due to the same cause. At the beginning of the solar system, an asteroid slammed into Venus at such an angle and with such ferocity that the direction of Venus' revolution was reversed and Venus wound up with no moon of its own. Theoretically, this could be undone by deflecting another asteroid and training it toward Venus, so as to reverse and speed up Venus' revolution and also to place the object in an orbit around the planet to serve as its Moon. Mind you, I said *theoretically!* Terra-forming of Venus would necessarily entail severe reduction in the dense cloud layers surrounding that planet. The entire beauty of Venus, in the eyes of us earthlings, is due to its very high albedo (reflectivity to sunlight). Reduction of the clouds will cause the planet to cease shining as bright as it does. Stripped of that pretty but poisonous cloud cover would make that planet appear not much more brilliant than mercury or mars.

The fundamental requirement for *terra-forming* Venus is the presence of a carbon cycle and there can be no carbon cycle on Venus because there are no oceans to be found there. The ingredients for prodigious amounts of water – hydrogen and oxygen –are present in abundance on Venus, but currently there is no way to wrest them from other compounds and make them into life-giving water.

Again, there are theoretical solutions for these apparently insurmountable obstacles. However, the cooling alone of our sister planet will not give us a new world to colonize.

All of the above discussion arose from an initial enthusiasm for instigating a runaway cooling of that cauldron by placing an *Al-Battani* type of barrier between the Sun and Venus. Only further reflection engendered by the study of the role of the *carbon cycle* defused that enthusiasm. Honesty requires that my original purpose for delving into this type of fantasy be balanced by equal attention to the reality of conditions on Venus. It is quite evident, however, that if 'terra-forming' of Venus is ever to be attempted by man, the very first step will have to be its cooling by placement of a barrier between it and the sun. If nothing else, this discussion serves to underscore how blessed is the combination of all the beautiful attributes of our planet and how it is incumbent upon us humans to preserve that exquisite balance between all that is optimum for life's existence and propagation.

APPENDIX A

AL-BATTANI – A LITTLE REMEMBERED
GIANT OF THE 'DARK' AGES

Al-Battani, Latinized as A*lbategnius* (full name Abū Abd Allah Muhammad ibn Jābir ibn Sinān al-Raqqī al-Harrānī al–sābi' al-Battānī) was born in Harran in Northwestern Mesopotamia (now in Southeastern Turkey) in 858 A.D. He lived variously in Damascus, Syria and the then newly founded city of Samarra in Iraq where he did most of his work and where he died in 929 A.D.

So great and fundamental were his original contributions to astronomy and mathematics that in his *'divine Comedy'* Dante Alighieri, that most implacable foe of Islam, made a special exception for Al-Battani by exempting him from, not just Hell but also from purgatory, even as he acknowledged Al-Battani's Islamic allegiance. And no wonder. Read on. (What follows is largely, but not exclusively, a summary of the elegant and scholarly article of Willy Hartner in the *"Encyclopedia of Scientific Biographies."*)

The Vatican inspired *Gregorian* reform of the now universally accepted Christian calendar based its entire, and very ambitious and successful, enterprise by wholly accepting Al-Battani's edict that the solar year consisted of 365 days, 5 hours, 46 minutes and 24 seconds! This number, first arrived at over 1,100 years ago has hardly been refined and promises to keep our calendar in synch for at least 40,000 more years.[49] The 40 years of observations, painstaking measurements and exquisite analysis of the data that lay behind the derivation of this precise figure were wholly original to Al-Battani. He also defined and measured the *sidereal* year and clarified the difference between it and the *solar* year. He measured the inclination of the Earth's axis relative to the plane of the ecliptic to the uncanny accuracy of 23^0 35' and used it to explain the seasonal variations in the lengths of day and night and the seasons themselves.

But dabbling with the calendar reform was only a side show for Al-Battani. His father, also a Muslim, but the descendent of people of a monotheistic cult of the *Sabeans* of Mesopotamia, antecedent and analogous to the Islamic faith, was also an astronomer/astrologer and maker of astronomical instruments and thus the interest in the stars ran in Al-Battani's family. Indeed, some people derogatorily ascribe to Sabeans the epithet of *star-worshippers,* an unfair appellation because the Sabeans, strictly monotheistic in their beliefs and practices, only maintained that the stars in the heavens were angels (later referred to in the Quran). And no wonder there because the Quran states that God created angels from *'smokeless fire'*. What is in the thermo-nuclear fusion furnaces of the interior of the stars if not *'smokeless fire'*?

In order to make sense of his four-decade-long observations which he catalogued in the immortal treatise with the unassuming title of 'The Sabean Tables" (*al Zij asSabi*[50]), Al-Battani invented,

---

49   According to Duncan Steel ('Marking Time'), if we wanted to extend the useful life of the Gregorian calendar to, say, a million years, we will have to replace it with the calendar proposed by another Muslim genius Omar Khayyam who propounded it about a hundred years after al Battani, living in the same general area of Middle East, at the confluence of Arabic-Persian-Turkish civilizations.

50   The title of this treatise, the manuscript of which is still extant, may or may not be pregnant with significance. I have always found it intriguing that many Western historians and chroniclers, not to mention the inimitable Dante, unreservedly express their adulation for this giant in

from scratch, the altogether new sciences of trigonometry and pre-calculus. Impatient with the then extant wordiness of mathematical theorems, Al-Battani introduced many of the signs for mathematical operators, thus vastly simplifying enunciation of mathematical equations and concepts. That Newton and Leibnitz published their text-books of the 'newly invented' science of calculus in the same year and with the same number of chapters in the same orders is not an enigma. Given the coincidence, it has a simple explanation. They had apparently both perused – and borrowed from – the recently translated and published mathematical works of Al-Battani and promoted them as their own original contributions. Newton did say that he was able to see farther than anyone else because he sat upon the shoulders of giants; he simply did not mention those giants by name!

Al-Battani greatly improved upon Ptolemy's measurements of the Sun's *apogee and* perigee and calculated the eccentricity

---

glowing terms, while Muslim writers seem to be studiously avoiding any mention of al Battani and, at best, seem bent upon marginalizing him. The reference to the Sabean connection may offer the clue. A few facts are germane here. (1) Not only Al-Battani himself but, without a doubt, his father, and perhaps his grandfather were also Muslims and pretty devout ones at that. (2) Sabean beliefs, unquestionably monotheistic in every sense, predated the advent of Islam and were more than tolerated by the Prophet Muhammad. (3) For that reason, perhaps, Sabeans may have been slow to embrace Islam formally. (4) More importantly, they probably saw no reason to repudiate their astronomical expertise which had provided the basis for the extra-ordinary navigational skill of Arab traders, both in the vast deserts of Arabia as well as on the high seas both North and South of the Equator. (5) By the ninth century A.D. the temper of the times had changed in the Islamic world and the erstwhile Sabeans were then being impugned as "star worshippers," a pejorative epithet among the champions of monotheism. (6) The title of the book suggests that Al-Battani may have built upon the legacy of his forebears as he refined their notes in his definitive treatise and felt compelled to acknowledge his debt to them, something that did not sit well with the demagogues who may have wrested the Islamic priesthood by that time.

of the Earth's orbit to values that are very nearly identical with our modern, very precise numbers. Likewise, his measurements of the rate of the precession of equinoxes leave no room for further refinement. He predicted lunar and solar eclipses and explained the astronomical bases of those phenomena and clearly enunciated celestial laws that were later to be named as Keppler's laws, Copernican theory and the three laws of motions credited to Isaac Newton more than five centuries hence. These facts clearly indicate that the heliocentric theory pertaining to the solar system was common knowledge among the Muslim astronomers during the tenth century AD and, perhaps, for many generations prior to that, antedating Copernicus by seven or eight centuries.

To his credit, Copernicus did profusely acknowledge his debt to Al-Battani, as did Tycho Brahe and Johannes Keppler; Newton and Leibnitz never mentioned him or any other Arab/Muslim scholar. Keppler most probably relied more on Tycho Brahe's tables than on Al Zij. Tycho, on the other hand, very definitely possessed in his library the Latin translations of several of Al-Battani's works and may simply be revalidating and confirming observation made seven centuries earlier to see whether the heavens had shifted since the times of Al-Battani.

Without a doubt, Al-Battani's work on *parallax*, his development of the essential mathematical tools of trigonometry and calculus, his cataloguing and clarification of the movements of the heavenly bodies, the enunciation of the laws underpinning those movements, and his prescient understanding of the nature of *smokeless fire* were stepping stones to the *relativistic* and *space age*. His extensive commentary of Ptolemy's (Al Batlamiyus, in Arabic rendering) *Tetrabiblos* (*Kitab al-arba'a*) seems effectively to put to rest Ptolemy's tortured and convoluted theories of epicycles to explain celestial movements.

Unlike the European scholars who drew their inspiration from the work of this master but did not acknowledge their debts to him, Al-Battani very matter-of-factly credited many of his assertions as being based on the Hindu scholars Aryabhatta and Gargi. Some degree of his vain-glorious trait is revealed in his disregard of his older contemporary Thabit Al-Qurra's exposition of the theory of *trepidation* and thus he incorrectly described the rate of the precession of equinoxes as being uniform (which it is not).

APPENDIX B

THE STORY OF THE ATOM

The period in human history and in the history of science spanning a mere half century between 1895 and 1945 was probably the most fecund epoch. The implications of the dazzling parade of scientific discoveries during that period for human destiny, indeed for the very fate of *Homo sapiens,* is, of course, not yet completely clear but will no doubt be monumental, whatever course humanity takes with regards to its natural impulses, whether peaceful or violent, good or evil.

Those discoveries, mostly in the field of physics, matched in the previous century by equally dazzling discoveries in the field of chemistry, spawned a great many new disciplines which now engage the brains and labors of an army of scientists and engineers numbering perhaps in the millions and financial resources of a coterie of nations measured perhaps in hundreds of billions of dollars. Aside from a full clarification of the source of heat energy emanating from the belly of our planet, the energy that underpins the blessed climate that makes the genesis and maintenance of all life possible, the understanding of the phenomenon of nuclear fission is a most intellectually satisfying exercise for the perceptive readers to engage in even as we concede that an exposition of the subject this detailed is probably very tangential to the main focus of the book.

Below I will try to cover the salient mileposts of a long series of discoveries and hypotheses that constitute homage to the glorious constellation of scientists and their achievements. This appendix on the story of the atom is almost entirely an encapsulation of the brilliant chronicles of the particle physics admirably narrated by Richard Rhodes in two massive – but eminently readable – volumes entitled "The Making of the Atomic Bomb" (1986) and "Dark Sun – The Making of the Hydrogen Bomb" (1995) published by Simon & Schuster.

Fifth Century B.C.: Thracian *Democritus* built on the speculations of *Leucippus* about atom as the reality in the nethermost world of matter. The original works of these giants have been lost but their claims to immortality were propagated by allusion to them by later Greek writers, Aristotle specifically, whose works, in turn, only survived in Arabic translations thanks to the early Muslim love of learning and intellectual speculation. Seventy-two books of knowledge alleged to have been written by *Democritus* and countless other Greek and other scholars are believed to have been lost when zealous Christian mobs burned libraries everywhere during the third through the fifth centuries because 'natural philosophy' was taken as the work of pagans and idolaters and, supposedly, its perusal was bound to weaken their Christian faith! The books – rare manuscripts of masters of the previous eight centuries, extant only in single copies – were burned in religious frenzies all over Europe, North Africa, and Asia Minor. The resultant Dark ages that descended upon Europe and so blighted Christendom were mitigated – feebly and for a brief period of only three or four centuries – by a flash of enlightenment in the Muslim world which itself fell victim to similar religious bigotry directed against secular learning.

In the subsequent centuries, the discussion of the atom was at the extreme fringes of '*Natural Philosophy*', a term which denoted what we today call *science*, because no intellectual pursuit or practical necessity compelled serious consideration of what an atom might be. As late as the early 20th century, giants like

91

Max Planck pooh-poohed the concept of the atom as a physical quantity because it collided with his dear laws of thermodynamics! This attitude of European physicists – of whom Planck can be considered archetypical – was curious, given the advances in chemistry that had been occurring at breakneck speed during the previous one hundred years and which clearly pointed to the existence of atoms as physical entities.

The period of drought in intellectual inquiry–started even before the decline of Greece as a major center of civilization shortly after the time of Alexander the Great and briefly interrupted with the ascendance of Muslim spark of learning from the 8th to the 12th centuries – has been variously attributed to the Roman disdain for anything not related to military, law, or public works, to climatic change, to pestilence, to the prevalence of feudalism and to the rise of commercialism. Edward Gibbon, in his seminal "History of the Decline and Fall of the Roman Empire" seems to ascribe the intellectual decay of the period commonly called the Dark Ages to the ascendance of Christianity and the support it received from the House of Flavians, its royal patrons. Be that as it may, until the dawn of the 13th century, all things intellectual were discouraged and persecuted by the Vatican until the onset of the Italian renaissance.

1807: John Dalton, in the city of Manchester, England, resurrected the Leucippus/Democritus theory of atoms.

1879: The incomparable *James Clerk Maxwell*, to his untimely death due to cancer, would not hear of the physical nature (physical in the sense of being material) of the atom. His admirable – and I might say perfect – theory of electromagnetism satisfied all the requirements of scientific arguments.

1894: Even at this late date, the Scottish giant of a scientist, Robert Cecil, in his Presidential Address to the British Association, in an amused and poignant but arcane prose dismissed the possibility of human beings ever elucidating the nature of the atom!

1895: *Wilhelm Conrad Röentgen* discovered X Radiation, at the high frequency extreme of the electromagnetic spectrum. *James*

Clerk Maxwell's studies, in the previous decades, had predicted the existence of such ultra small wavelength (fifty thousandth as small as the smallest wavelength in the visible light spectrum) a mere two years before *Röentgen's* discovery. Heinrich Rudolf Hertz had described the *radio waves*, intermediate in wavelength (and frequency) between visible light and X Rays and *Röentgen* was quick to see his discovery as further extension of the theoretically limitless electromagnetic spectrum. However, X Rays had some extraordinary and unique properties, not the least of which was their ability to penetrate solid objects and their propensity to induce *ionization*[51] of the medium through which they were traveling. Soon thereafter, Max von Laue used X Rays to study the internal structure of crystalline chemicals and, Lo and behold, man could now see the atomic lattice whereupon all doubts about existence of atoms should have been extinguished.

1896: a few short weeks after reading *Röentgen's* paper, Frenchman Henri Becquerel, member of a dynasty of physicists ruling over the French *Musée d'Histoire Naturelle* in Paris, demonstrated that Uranium salts spontaneously emitted radiation that exposed photographic plate in much the same way as did X Rays.

1897: J.J. Thompson, the director of Cavendish Laboratories at Cambridge University announced the discovery of the *electron* and further demonstrated that these beams, emanating from cathodes in sealed glass tubes emptied of all gas, could be bent by application of the magnetic field; therefore, he asserted, they were not waves but infinitesimally small particles,. He further showed that all electrons were identical in every respect no matter what their source. So now we have a part of something bigger and

---

[51]   In 1884, Svante August Arrhenius of Sweden had laid out the theory of ionization in his thesis as a doctoral candidate. It was dismissed as being unfit to qualify him for the Ph.D. but good enough to earn him the Nobel Prize some twenty years later! Michael Faraday, several decades earlier had postulated the phenomenon but only during the passage of electric currents. Faraday, however, deserves the credit for the coinage of the terms, ion and ionization.

that bigger something has to be the atom! Furthermore, what remained of the atom after the electron was stripped away was positively charged. Enter Ernest Rutherford, the great master's prodigious protégé. The first order of business for Rutherford was to show that the energetic spontaneous radiation emanating from uranium salts was, in fact, also made of a stream of electrons.

1899: Upon further study, Rutherford discovered that in addition to electrons (which he called the β radiation, which had greater penetrative power), uranium and thorium emitted another radiation with very limited ability to penetrate; he called it α radiation. Meanwhile Rutherford, at age twenty-seven had accepted the position of Professor of Physics at McGill University in Montreal from where he published his work.

1900: In collaboration with *Frederick Soddy*, a young chemistry genius, Rutherford discovered spontaneous transmutation of thorium (atomic number 90, atomic weight 232) into inert gas Argon (atomic number 18, atomic weight 40), the first time a *radioactive* element was observed to disintegrate spontaneously, an earth-shaking enough discovery in itself to earn Soddy a Nobel Prize. A byproduct of that research was the phenomenon of half-life of radioactive substances and existence of different *isotopes* (of the same element), a term coined by Soddy at a later date. Later that year Max Planck elucidated the concept of quanta and propounded Planck's constant, work that laid the foundation for Neils Bohr's work cutting the Gordian knot of the paradox concerning the stability of the atomic nucleus. (See below)

1903: In another stunning revelation, Rutherford and Soddy showed that the amount of energy released by disintegrating atoms was spooky, between 20,000 to one million times the amount released by any known *exothermic* chemical reaction! German physicist *Phillip Lenard* demonstrated the extent of the empty void inside the atom by bombarding elements with cathode rays.

A Japanese theoretical physicist *Hantaro Nagoaka* proposes a Saturnine model of the atom, with a positively charged nucleus surrounded by negatively charged particles orbiting around the former, analogous to Saturn and its rings, very close to the modern concept of atomic structure!

1904: Soddy clearly hinted at – but immediately recoiled from – the possibility of someone, some day releasing vast quantities of energy locked up in the atomic nuclei of heavy elements and putting them to evil use! Rutherford pronounces the idea as '*moonshine*'.

1905: The German chemist from Frankfurt *Otto Hahn*, in Montreal to work under Rutherford, showed that the alpha particles released from disintegrating thorium, uranium (atomic number 92, atomic weight 238) and actinium (atomic number 89, atomic weight 227) were indistinguishable from one another. Based on that, Rutherford, in **1908**, now back in England, made a definitive pronouncement that α particles were charged helium atoms.

Albert Einstein, commonly acknowledged to be the greatest genius the human species has ever produced, propounded his Special Theory of Relativity in an unassuming, poorly crafted paper published in *Annalen der Physik*. The science of physics – both theoretical and experimental – was now picking up tempo that would soon culminate in the most significant scientific event of the twentieth century, the harnessing by man of the prodigious quantities of energies locked up since the time of *creation;* the new physics would point to the source of evidently limitless energy within the bowels of earth and in the interior of stars. Not only does Einstein sew up the concept of the unity of matter and energy as two sides of the same coin but he establishes the mathematical construct to elucidate the formula ($E=mc^2$) denoting relation between the two.

1906: Rutherford, again at McGill, noticed the massive deflection of some of the energetic α particles while passing through an ultra-thin sheet of mica. Based on this he surmised the existence of immensely strong electrical forces within the nucleus. His calculations yielded the value of 100 MEV per square centimeter as the intensity of the electrical field, an unbelievable magnitude! In collaboration with his junior assistant *Hans Geiger*, the designer of <u>scintillation</u> <u>counter</u>, he conducted experiments that steered their research on alpha particles in the inevitable direction of defining and revealing the nucleus as the seat of most of the atom's weight in the center of an immense void.

1911: Based on the astonishing finding that some of the alpha particles bounced back off a thin gold foil, instead of penetrating it like the vast majority of the particles in the beam, in a series of experiments conducted by *Hans Geiger* and an 18 year old undergraduate *Ernest Marsden* during the preceding years, Rutherford was able to make a definite pronouncement about the nature of atomic nucleus: a very heavy but unbelievably small center within an empty space comparable in magnitude to a tennis ball in the center of a large stadium. However, great theoretical objections persisted, both to Nagoaka's saturnine model, based solely on speculation, and to Rutherford's model, derived from undeniable experimental data. Because of electrical repulsion between negatively charged particles orbiting the nucleus, the whole structure should immediately tear apart but it doesn't. *C. T. R. Wilson's* invention of the *cloud chamber* a few months later made possible the actual visualization of bouncing alpha particles with photographs, unaided by microscope.

1913: *Niels Bohr,* the 29-year-old Danish giant of a physicist, ranked right along with *Einstein* in the highest echelon in the pantheon of scientists, exploded on the world of physics with his landmark paper. Departing from time-honored principles of classical (Newtonian) mechanics because it could not account for the extraordinary stability of the atomic nucleus, Bohr sought, and received, the answer in Planck's postulate concerning quanta.

1915: *Albert Einstein* publishes his 'General Theory of Relativity' and turns the world of the theoretical physics – especially Newtonian Mechanics –on its head and spawns generations of serious scientists, pseudo-scientists, and science fiction writers.

Many of the latter are utterly incapable of understanding the rather difficult concept propounded by this genius.

1919: Rutherford, meanwhile, following up on Marsden's improbable findings in 1915, achieves release of the 'hydrogen atom' from nitrogen by bombarding the latter within a scintillation chamber with alpha particles (helium atoms), sourced from radium (88, 226), the first *transmutation* of an element by man; a true alchemical achievement (!). At the time, it was mistakenly hailed as '*splitting*' of an atom of nitrogen but later analysis clarified the phenomenon. What had really happened was that one in 30,000 nitrogen atoms (atomic weight 14) had been transmuted into ^{17}O isotope of Oxygen and reduced the helium atom to a naked hydrogen (which Rutherford promptly christened a *proton)*.

In the train of this experiment, Rutherford and others working independently achieved transmutation of many lighter elements but found that the heavier elements resisted transmutation because of increasingly tough electrical barriers in their nuclei. The technology to achieve the acceleration of particles was not available yet and the progress of nuclear physics stalled.

1920: *Francis William Aston*, working with J.J. Thompson at the Cavendish Laboratories in Cambridge devised the *mass spectrograph*. The instrument he devised utilized serial separation of different nuclei by passing the mixture of many elements successively through electrostatic and then magnetic fields which blackened a strip of film in a characteristic array enabling him not only to identify individual elements but also their various isotopes. Bolstering Rutherford's model of atomic nuclear structure was a welcome byproduct of Aston's work. It enabled Aston to identify 212 out of 281 naturally occurring isotopes of the 92 elements in the periodic table. He also convincingly demonstrated that the weights of every isotope he studied were *whole numbers.* What this meant was that each element was assembled simply from protons, neutrons and electrons – the true building blocks of the universe!

Explaining the minute but persistent and significant departures from whole number scenario, especially with regard to hydrogen and helium atoms, Aston proposed the existence of a glue called *binding energy,* a concept very consistent with Einstein's equivalence of mass and energy and foretelling the possibilities, first of unlocking the binding energy into harnessable power and, next, of outright transformation of mass into energy, prodigious amounts of it.

At the Bakerian Lecture before the Royal society, Rutherford openly muses about the *neutron,* confidently speculating about its properties and its potential use for further probing the atomic nucleus. His belief, even then, was that the neutron is not an independent *third* elementary particle but probably consisted of the positively charged proton and negatively charged electron, somehow bound so tightly as to act as one particle. This refusal to fully accept the neutron as a separate particle chagrins his protégé and close associate Chadwick who, unfortunately, had lost out several years of work having been imprisoned in the German internment camp during the Great War. Chadwick was still convinced about the existence of the neutron as an independent particle, based on the demonstrated disparity between the atomic number and atomic weights of almost all the elements. See below under **<1932>**

The *Great War* in part accounts for the relative stagnation in the march of nuclear physics between 1915 and the late twenties, not least with the recruiting of many of the best minds of the warring countries to serve military ends. The hiatus in imagination was undoubtedly manifested in good part by the difficulties and costs of developing particle accelerators. It was the ushering in of the development of *Big Machines* in America that gave impetus to the resumption of the forward march of physics, despite the dislocations caused by the harassment and expulsion by the Nazi regime of resurgent Germany of many Jewish scientists, most of whom migrated to avoid persecution but also in search for opportunities to pursue their calling. By

1930 the tempo of research was again picking up, this time with American scientists joining the fray in earnest.

1931: *Ernest Orlando Lawrence* and his student *M. Stanley Livingston* unveil the first functioning *cyclotron*, a unique and imaginative adaptation of the concept of a linear accelerator at the University of California at Berkeley. Particle accelerating powers in the range of millions of electron volts had now become possible.

1932: *James Chadwick* of the Cavendish laboratory of Cambridge University, on February 27 postulated the existence of the *neutron* and four months later positively demonstrated its existence. He asserted that all neutrons were the same, no matter which elements they came from. The significance of this epochal discovery is impossible to exaggerate. All sustained (controlled or uncontrolled) nuclear fissions are possible only when fast moving neutron strike nuclei of (heavy) elements, resulting in the release of energy trapped as atomic binding energy, in prodigious amounts bordering on the confluence of matter and energy. (Ironically, in his 1914 novel, *World Set Free!,* the great H.G. Wells uncannily predicted man's harnessing – for good as well as evil – atomic energy; only he picked the year for the cataclysmic war between the allies and the central powers to be 1956 when all the major population centers in the Western civilization would go up in atomic flames).

More importantly, because neutrons, having no positive charge do not get repelled, the discovery of the neutron completely eliminated the need for accelerating the protons to monstrous speeds to overcome the electrical repulsion between the positively charged protons in the nucleus and the bombarding protons. Eventually uses were found for cyclotron which were designed to accelerate alpha particles (protons), but not for bombarding the nuclei to cause them to break apart.

1933: The Hungarian physicist *Leo Szilard*, while an anonymous refugee from Nazi Germany in London, fancied the scenario of the neutron bombardment of some – as yet unidentified –

sufficiently heavy and unstable element to start a nuclear chain reaction. It was *Enrico Fermi*, in **1942**, who actually accomplished such bombardment to achieve a controlled nuclear chain reaction in the basement Gym of the University of Chicago. The *atomic pile* was born.

At the seventh Solway Conference held in Brussels, the American Carl Anderson, from Caltech, postulated the existence of antimatter. Anderson announced his discovery (made the year before during *cosmic ray* research) of the positron as one of the elementary building blocks of matter.(The positron was found to have the same mass as the electron, but carried a positive charge).

Enrico Fermi while heading the physics department at the University of Rome published an important paper on *Beta decay*; this work remains the definitive theory on the subject. Almost a corollary to the phenomenon of beta decay, Fermi also postulated the existence of the *weak force*.

1934: Physicist couple *Irene Curie* and *Pierre Joliot,* stung by chastisements from colleagues at the Solway Conference, especially by *Lise Meitner,* first for missing the neutron and later the positron, went to work. Using bombardment by alpha particles, they achieved the creation of an artificially radioactive element, by transmutation of aluminum into radioactive isotope of phosphorus (with an extremely short half-life of three minutes) which they were deftly able to save up long enough for very fast chemical identification through imaginative maneuvers. This achievement won them the 1935 Nobel Prize in chemistry. The important discovery demonstrated not only transmutation with artificial radioactivity but the release, in the process, of some of the atomic energy.

Fermi, in Rome, immediately saw the promise of Curie-Joliot's work and seized upon the usefulness of the neutron as the probing tool for atomic nucleus. Not discouraged by lack of any results of neutron bombardment (obtained from radon gas) of lighter elements, he first hit pay-dirt with aluminum and then, in rapid succession, with sequentially heavier metal and non-metals. Most of the artificially radioactive products had a short half life, sometimes no more than a few minutes, and so his team had to work skillfully and nimbly by running the distance from one end of the lab to another to avoid exposing the new elements to contamination with stray radiation. When they reached the heavier elements, they found that the bombarding neutrons just got added to the nuclei of, say, uranium; the resulting product was a heavier isotope of the original metal. The energy of the fast moving neutron simply became incorporated in the binding energy of the new isotope. In the case of uranium, the artificial isotope (239) decayed into a new man-made element with the atomic number of 93. Although it only had a fleeting existence, it was nevertheless a transuranic element, something which does not exist in nature. Emission of a Gamma photon was neatly causing the neutron to be amalgamated into the nucleus of the new element as a proton! Segre and Amaldi coined the phrase *radiative capture* for this nearly magical happening.

June 28 and July 4: To an amendment to his earlier patent application, Szilard added the phrase "Liberation of nuclear energy for power production and *other purposes* through nuclear transmutation" (emphasis mine). Also, he proposed the concepts of *chain reaction* and *critical mass* and prophetically bragged about his ability to create an *explosion* using tamper around a fissionable element (ultra light Beryllium, not ultra heavy Uranium according to his reckoning!) of critical mass arranged in a sphere. Marie Curies dies on that same Fourth of July!

1935: Certain about the imminent release of huge atomic energy and its Inevitable military use by the 'bad guys', Szilard surrenders his patents concerning nuclear chain reactions and explosions to the British admiralty on condition that they be kept secret. .

1936: In a landmark lecture before the Danish Academy, *Niels Bohr* poured cold water on the idea of releasing atomic energy even by bombarding very heavy nuclei with multi-million electron volt neutrons because the neutrons simply got captured and the subject nucleus simply became its heavier isotope or an element higher up the periodic table, releasing only gamma photons in a '*beta decay*'. This was really a reiteration of Rutherford's assertion decades earlier that the idea of releasing nuclear energy was '*moonshine*'! Einstein likewise had dismissed the idea. Bohr reasserted, with added mathematical argument, this denial in a paper published the following year. All the major figures in the world of theoretical physics – including Frisch, Meitner and others – embraced his views; such was the reputation and stature of this giant! That year, 1937, also marked the passing away of Rutherford. But the <u>punch line</u> was to come soon, in fact a little over a year later in _____

1938: *Otto Hahn,* with help from his young assistant *Fritz Strassmann,* working at the Kaiser Wilhelm Institute for Chemistry, the old haunt of now emigrated Lisa Meitner, in the Berlin suburb of Dahlem. Hahn and Strassmann are unsuccessful in efforts to separate the supposed newly created heavy radioisotope (from bombardment of uranium with neutrons) from barium used as a carrier in chemical fractionation, despite endless and very careful repetition of experiments under the most stringent conditions. The duo is forced to conclude that the daughter nuclide was emphatically not the expected heavyweight offspring Radium III but the lowly Barium (element 56), removed 40+ places down the periodic table and almost half as heavy as the parent nuclide!

In desperation, Hahn turned to Meitner, now marooned in Sweden, jobless and despondent but not by any means robbed of her gift of scientific alacrity. While awaiting Meitner's response, the duo again ran the experiment and was left with no doubt that the daughter nuclide was Barium which had been bumped, through ejection of the gamma photon, one place up the periodic table to the lanthanum (element 57)! A new epoch had broken out because now it was not only transmutation with loss of a proton or even an alpha particle; the actual shattering of a heavy nuclide had become possible! Meitner meanwhile was in the resort village of Kungalv, a guest along with her nephew

Otto Frisch, now an émigré to Copenhagen of some friends for Christmas dinner. While the nephew was in Sweden for only a couple of days, the two knocked their heads together trying to make sense of Hahn-Strassmann experiments at KWI. Recalling Bohr's depiction of the nucleus as being analogous to a liquid drop, and surmising that the "surface tension" (really the *strong force)* holding together that heavy a nucleus must be very tenuous. They postulated, correctly, that the fission of the uranium nucleus into two roughly equal fragments (barium and krypton in this case) must therefore be possible. They conceded that that is exactly what was happening at KWI. But lo and behold, further calculations unfolded a scenario for which neither Meitner and Frisch nor Hahn and Strassmann, were prepared; nor, for that matter, was anyone else in the physics community. Even the stubborn Szilard had despaired of the possibility! So here is the punch line: **Meitner and Frisch's *calculations showed that the fission of just one atom must yield 200MeV, a prodigious and frightening quantity of energy.*** The rest, as they say, is history.

When Leo Szilard (by then lodged in the US) read the landmark paper by Hahn and Strassmann, in January, 1939, he went ballistic. He cornered Albert Einstein and prevailed upon him to accept the grave significance of the Hahn/Strassmann findings and their interpretation by Meitner and Frisch. Szilard emphasized to Einstein the real possibility that Hitler may use the new discoveries to embark on a crash program of developing an atom bomb. Reluctantly, Einstein relented. Yielding to the pressure from Szilard (whom he had so far regarded as nothing but a crackpot!), in early August of 1939 Einstein signed the first of his famous "Dear Mr. President" letters (largely written by Sziilard) to Franklin Delano Roosevelt urging him to immediately marshal all the resources of the United States to the enterprise that came to be known as the Manhattan Project.

APPENDIX C

THE CASE AGAINST SOLAR VARIABILITY
FOR EXPLAINING CLIMATE FLIP-FLOPS

I am devoting this Appendix of substantial size to a subject which is quite controversial and perhaps completely dismissed as irrelevant by the consensus of serious climate scientists; this is because we might find ourselves, at least in the beginning, outside the pale of credibility, what with our advancing somewhat unconventional ideas and our reluctance to accept the majority of these assertions as if they were revealed truth. Our very competence to dabble in the issue of global warming may be subject to challenge. Those who advocate the pivotal role for the ups and downs of the solar activity in the causation of the global climate flip flops, on the other hand, seem to me to be scholars with substance and good standing.

Here, in a nutshell, is the contention. The energy output of the Sun varies from time to time; the periodicity of such variations, while not evidently rhythmically regular, nevertheless follows a pattern which coincides directly with the advent of warm and cold epochs on our planet. It is further contended that the richness or poverty of the crops of sunspots is a harbinger of the ferocity of the Sun's energy output. We need to be familiar with a few phenomena related to solar activity about which there is not much debate.

Sunspots and sunspot cycles: Sunspots[52] are cool regions on the sun where magnetic energy builds up – about 1,000 to 1,500 degrees cooler than the surrounding areas. The more intense magnetism prevailing in the regions of the sunspots interferes with convection currents of heat from the Sun's interior which explains their relative coolness. They serve as a cap on material welling up from below. Often, that material is released in spectacular light shows– solar flares and discharges – of charged particles known as coronal mass ejections. The ejections can travel as space storms to Earth within a day or so, and major storms can knock out satellites and trip power grids on the surface. *"All this matters because, as laid out in a report earlier this year by the National Academy of Sciences, a major solar storm nowadays could cause up to $2 trillion in initial damages by crippling communications on Earth, fueling chaos among residents and even among governments in a scenario that would require four to ten years for recovery. Such a storm struck in 1859, knocking out telegraph communications and causing those lines to erupt in flames. The world then was not so dependent on electronic communication systems, however."*

Sunspots are several thousand kilometers across in their greatest extent and appear in crops, close to the poles of that star and persist for several months to a few years before fading. They rotate in synch with the rotation of the Sun.

The abundance of sunspots follows a fairly regular cycle of roughly eleven years, the so-called sunspot cycle. *Solar Maxima* denote the periods of maximum number of sunspots during the cycle; conversely, *Solar Minima* denote a period of paucity.

---

[52]   Galileo Galilei, in 1612 definitively pronounced existence of sunspots (first observed by him in 1610 and soon thereafter by several astronomers). Hitherto Sun, God's perfect creation, was considered free from any 'blemishes' and Galileo's pronouncement had the Vatican up in arms against this unregenerate heretic. Harsh denunciations of all of Galileo's work started building up and culminated in his denunciation by the Roman Inquisition and eventual house arrest for the rest of his considerable remaining life

The adherents of sunspot theory do not contend that the global climate closely follows the sunspot cycle, firstly because the overall output of the Sun's energy is not diminished with the abundance of sunspots (the areas of the photosphere in the immediate vicinity of sunspots actually get hotter and cancel out the lower radiant output corresponding to the sunspots) and secondly, because no cycle of climate change has been documented that would dovetail with sunspot activity.

Maunder Minimum, Sporer Minimum, Dalton Minimum and other harbingers of '*Little Ice Ages*'

The bedrock arguments in favor of correlation between sunspot activity and global climate upheavals are based on the observation related to these minima during which the number of sunspots fell to exceptionally low numbers over prolonged periods during which the global climate (or the climate in the Northern Hemisphere at any rate) experienced marked cooling over an equally long period, the two events coinciding uncannily with each other.

The facts are indeed striking. In two papers published in 1890 and 1894, respectively, the renowned English solar astronomer Edward Walter Maunder (1851-1928) unearthed the observations made contemporaneously and documented with care concerning sunspot activity during 1645-1715, a period universally documented as a "Little Ice Age." So intense and sustained was the cold spell that, for instance, people could walk from Manhattan to Staten Island over New York's frozen outer harbor in late spring! Stories of a similar nature reported for conditions in Europe abound. Whereas in recent times, 600 to 700 sunspots are observed in a typical year, the observed numbers fell on average to less than one a year during that Little Ice Age. That specific period is officially known as the *Maunder Minimum*. Sunspot cataloguing for earlier such Little Ice Ages was not made and is not available directly but has been surmised indirectly by the Radiocarbon method and one such minimum, the so-called *Sporer Minimum* (1450-1540 AD), named after

German astronomer Gustav Sporer (1822-1895) seems to hold up. Less convincing are the data about the shorter (1790 to 1820 AD) *Dalton Minimum* named after English meteorologist John Dalton[53] (1766-1844) during which, besides very subdued sunspot activity, a series of four super-volcano eruptions occurred between 1812 and 1815 and the enormous amount of volcanic ash that overspread the whole globe can fairly explain the period of intense cold.

The paradox that the proponents of this thesis are unable to explain is this: A sunspot is a cool spot. The paucity of sunspots[54] over a sustained period should boost insolation and should cause global warming and not a Little Ice Age. Could it be that the photosphere overcompensates for the cooler sunspots when they are plentiful and may back off during their period of scarcity? That is a very anthropocentric viewpoint and serious science should not entertain such cupidity, in my opinion. On the other hand, it has been noted that solar output actually increases during high sunspot activity and decreases during the periods of its paucity. The correlation has been inconstant and only tentative in recent times. For another divergent view please visit the following link in "Sky& telescopes web-site: http://www.skyandtelescope.com/ community/skyblog/newsblog/56631092.html

---

[53] The same John Dalton who resurrected the atomic theory of the ancients.

[54] Robert Roy Britt of Space.com reported on July 6, 2009 about the recent end of a two year long paucity in sunspot activity. According to the piece cited, during the preceding two years, the sunspot activity was completely absent for 77% of the time and that the current drought is the most pronounced since 1913. There has been no current cooling of the global climate during this period.

APPENDIX D

KYOTO PROTOCOL AND COPENAGEN
CONFERENCE ON GLOBAL CLIMATE

Below is the précis of the so-called Kyoto Protocol adopted in 1997 by all the countries of the world. It has been ratified by every signatory government except the United States. This précis has been published by the Directorate General of the European commission on environment: http://ec.europa.eu/environment/contact/contact_en.htm

The Kyoto Protocol

The Kyoto Protocol to the United Nations Framework Convention on Climate Change strengthens the international response to climate change. Adopted by consensus at the third session of the Conference of the Parties (COP3) in December 1997, it contains legally binding emissions targets for Annex I (developed) countries for the post-2000 period. The EU and its Member States ratified the Kyoto Protocol in late May 2002.

By arresting and reversing the upward trend in greenhouse gas emissions that started in these countries 150 years ago, the Protocol promises to move the international community one step closer to achieving the Convention's ultimate objective of preventing "dangerous anthropogenic [man-made] interference with the climate system." The developed countries commit themselves to reducing their collective emissions of six key greenhouse gases by at least 5%. This group target will be achieved through cuts of 8% by Switzerland, most Central and East European states, and the European Union (the EU will meet its target by distributing different rates among its member states); 7% by the US; and 6% by Canada, Hungary, Japan, and Poland. Russia, New Zealand, and Ukraine are to stabilize their emissions, while Norway may increase emissions by up to 1%, Australia by up to 8%, and Iceland 10%. The six gases are to be combined in a "basket," with reductions in individual gases translated into "CO_2 equivalents" that are then added up to produce a single figure.

Each country's emissions target must be achieved by the period 2008-2012. It will be calculated as an average over the five years. "Demonstrable progress" towards meeting the target must be made by 2005. Cuts in the three most important gases – carbon dioxide (CO_2), methane (CH_4), and nitrous oxide (N_2O) - will be measured against a base year of 1990 (with exceptions for some countries with economies in transition).

Cuts in three long-lived industrial gases – hydrofluorocarbons (HFCs), perfluorocarbons (PFCs), and sulphur hexafluoride (SF_6) - can be measured against either a 1990 or 1995 baseline. (A major group of industrial gases, chlorofluorocarbons, or CFCs, are dealt with under the 1987 Montreal Protocol on Substances that Deplete the Ozone Layer.)

Actual emission reductions will be much larger than 5%. Compared with emissions levels projected for the year 2000, the richest industrialized countries (OECD members) will need to reduce their collective output by about 10%. This is because many of these countries will not succeed in meeting their earlier non-binding aim of returning emissions to 1990 levels by the year 2000; their emissions have in fact risen since 1990. While the countries with economies in transition have experienced falling emissions since 1990, this trend is now reversing.

Therefore, for the developed countries as a whole, the 5% Protocol target represents an actual cut of around 20% when compared with the emissions levels that are projected for 2010 if no emissions-control measures are adopted.

Countries have a certain degree of flexibility in how they make and measure their emissions reductions. In particular, an international "emissions trading" regime is established allowing industrialized countries to buy and sell emissions credits amongst themselves. They will also be able to acquire "emission reduction units" by financing certain kinds of projects in other developed countries through a mechanism known as Joint Implementation. In addition, a "Clean Development Mechanism" for promoting sustainable development enables industrialized countries to finance emissions-reduction projects in developing countries and receive credit for doing so.

They pursue emissions cuts in a wide range of economic sectors. The Protocol encourages governments to cooperate with one another, improve energy efficiency, reform the energy and transportation sectors, promote renewable forms of energy, phase out inappropriate fiscal measures and market imperfections, limit methane emissions from waste management and energy systems, and protect forests and other carbon "sinks".

The Protocol advances the implementation of existing commitments by all countries. Under the Convention, both developed and developing countries agree to take measures to limit emissions and promote adaptation to future climate change impacts; submit information on their national climate change programs and inventories; promote technology transfer; cooperate on scientific and technical research; and promote public awareness, education, and training. The Protocol also reiterates the need to provide "new and additional" financial resources to meet the "agreed full costs" incurred by developing countries in carrying out these commitments.

The Conference of the Parties (COP) of the Convention also serves as the meeting of the Parties (MOP) for the Protocol. This structure has been established to facilitate the management of the intergovernmental process. Parties to the Convention that are not Parties to the Protocol will be able to participate in Protocol-related meetings as observers.

The agreement is being reviewed. The Parties will take "appropriate action" on the basis of the best available scientific, technical, and socio-economic information. Talks on commitments for the post-2012 period are on-going [see UN Climate Change Conference 2007, 3rd Meeting of Parties (COP/MOP-3) to the Kyoto Protocol, Bali]

For more information about and to peruse the text of the Agreement please visit:

http://en.wikipedia.org/wiki/Kyoto_Protocol\

The mandate of the Kyoto Protocol will run out in 2012 and it is expected that it will be supplanted by another, more definitive treaty. The *United Nations Framework Convention on Climate change (UNFCCC),* in cooperation with Intergovernmental Panel on Climate Change, will assemble Dec. 6-12 in Copenhagen, Denmark to complete the installation of a new treaty to replace the Kyoto Protocol. For More information about it and the 2007 Bali Conference visit: http://www.erantis.com/events/denmark/copenhagen/climate-conference-2009/index.htm.

Most of the contents of this appendix are quoted from Wikipedia (No permission required).

APPENDIX E

OTHER NOVEL PROPOSALS FOR
INTERCEPTING SOLAR RADIATION

There have been several suggestions for mechanically blocking the Sun's radiation sufficient to reduce it by more than 1 percent but less than 2 percent. Professional meteorologists estimate that curtailment at that level could reduce Earth's temperature by 1.75 degrees centigrade, enough to cancel out the expected rise in the 'global' temperature over the next half century. This degree of cooling is in line with the goals recently set by the G8 nations (The group of eight most industrialized nations, with 'free market economies', of course) held in July 2009 in L'Aquila, Italy. As currently envisioned by the international community, the route to achieving that goal would, of course, be by cutting CO_2 emissions[55]

One particular proposal by Jerome Pearson and colleagues as recounted in their 2005 article[56] was studied in detail because of the concrete figures they provide and the rather extensive set of references and summary of the literature.

---

[55]   As reported by Richard Harris on Morning Edition of National Public Radio News, July 9, 2009

[56]   See #2 in the bibliography

Briefly, Pearson and colleagues suggest creation of an artificial planetary ring around the earth, similar to the B ring surrounding Saturn (and some other gaseous giants like Uranus), but much closer. These rings, consisting of passive particles or controlled spacecraft with parasols, measuring as much as 1.2 to 1.6 earth radii (9,500 to 11,200 miles across) would shade the tropics. Use of materials obtained from Earth, Moon or asteroids for construction of such rings is proposed. They estimate the cost of the space ring at between 6 and 200 trillion dollars.

The costs alone would seem to put the project beyond our capacity to accomplish this in any foreseeable future. However, one feature of the rings, namely enormously increased illumination of earth during the nights might be even more problematic. An additional drawback of the space ring is acknowledged to be its interference with artificial satellites and spacecraft orbiting in low earth orbits. Another problem cited by the authors would be the irreversibility of the space ring. The cost of controlled spacecrafts with reflectors, on the other hand, is estimated at between much more manageable sums of $125 to $500 billion.

Astronomer Robert Angel[57] proposes a similar concept, a ring-like cloud of 'reflectors' circling the earth a million miles out. The objections to this concept are going to be the same as recounted above except that his cost estimates are several fold higher than the solution suggested by Pearson, et al.

Pallab Ghosh' recent piece on BBC News[58] reported on additional new ideas that are being promoted.

Among a number of articles in a similar vein, cited by the authors, a few antedate the April 1998 letter to the Editor of *the Atlantic Monthly* commenting on Calvin's January 1998 article on *climate flip-flop* referred to elsewhere. The suggestion for

---

[57] As quoted by The National Geographic magazine, August 2009 ("Shading the Earth")

[58] Please go to http://news.bbc.co.uk/2/hi/science/nature/8231387.stm

placement of a *cum-sole* barrier between sun and earth was first made then.

Interestingly, many of the proposals, e.g., Early's suggestion[59] calls for placement of the barrier at the Lagrangian point L1. Elsewhere in this book we have recounted reasons why such placement is impractical and not likely to be helpful.

And finally, we must point out the implicit and coincidental rejection by Pearson and colleagues of human culpability (through increased CO_2 emissions) for global warming. It is quite clear, according to them, that global warming is an inevitable process as we continue our emergence from the Ice Age and that a rise in sea levels and flooding of coastal areas is unavoidable and that human activities since the advent of the industrial age have little or nothing to do with it.

---

[59]   J.T. Early, Space-based solar shield to offset greenhouse effect, Journal of the British Interplanetary Society 42 (1989), pp. 567-569

BIBLIOGRAPHY

1. Calvin, William "The Great Climate Flip Flop," Atlantic Monthly January 1998
2. J. Pearson, J. Oldson and E. Levin: Earth Rings for planetary environment control, Acta Astronautica (online) June 27, 2005
3. Scientific American January 1909
4. Willy Hartner: Encyclopedia of Scientific Biographies on Al-Battani, pp507-516
5. IPCC, 2007: Summary for Policymakers. In: *Climate Change 2007: The Physical Science Basis. Contribution of Working Group I to the Fourth Assessment Report of the Intergovernmental Panel on Climate Change* [Solomon, S., D. Qin, M. Manning, Z. Chen, M. Marquis, K.B. Avery, M.Tignor and H.L. Miller (eds.)]. Cambridge University Press, Cambridge, United Kingdom and New York, NY, USA.

Suggested Reading

(1) "Marking Time" by Duncan Steel, John Wiley & Sons, Inc. 1955

(2) "Making of the Atomic Bomb" by Richard Rhodes, Simon & Schuster, 1995(paperback)

(3) "Dark Sun – The making of the Hydrogen Bomb" by Richard Rhodes, Simon & Schuster 1996 (paperback)

(4) "Climate, Man and History" by Richard Claiborne. W.W. Norton & Company, Inc, New York 1970

(5) "An Inconvenient Truth" by Al Gore, Rodale Press, New York N.Y., 2006

Other references: http://www.npr.org/templates/story/story.php?storyId=106658672

For an account of BBC's Pallab Ghosh summary of other ideas on space engineering please visit this web link: http://news.bbc.co.uk/2/hi/science/nature/8231387.stm

And for a recent report on solar activity and climate change: http://www.skyandtelescope.com/community/skyblog/newsblog/56631092.html

INDEX

X